VOLUME 4

APPLIED PHYSICS AND ENGINEERING

An International Series

Sophisticated Signals and the Uncertainty Principle in Radar

Sophisticated Signals and the Uncertainty Principle in Radar

D. E. Vakman

Edited by
Ernest Jacobs

Translated by
K. N. Trirogoff

Aerospace Corporation
El Segundo, California

SPRINGER-VERLAG NEW YORK INC. • 1968

ISBN-13: 978-3-642-88215-9 e-ISBN-13: 978-3-642-88213-5
DOI: 10.1007/978-3-642-88213-5

 Library of Congress Catalog Card Number 68-18104.

Softcover reprint of the hardcover 1st edition 1968

Title No. 3894

Preface to the English Translation

This book is a translation of D. E. Vakman's "СЛОЖНЫЕ СИГНАЛЫ И ПРИНЦИП НЕОПРЕДЕЛЕННОСТИ В РАДИОЛОКАЦИИ", literally, "Complex Signals and the Uncertainty Principle in Radar Technology". The word "complex" in the title stands for signals whose product of duration and spectrum substantially exceeds unity. Since "complex" is common English terminology for a signal that has real and imaginary components, we have taken the liberty of substituting "sophisticated" for "complex" whenever reference is made to signals with large duration-spectrum products.

The attempt has been made to keep the translation as faithful as possible within the constraints of good English usage and in keeping with differences of the technical terminologies between English and Russian. Further, the redundancy so common in Russian writing was often times reduced so as to conform with the more succinct styles used in English. The English language editions have been substituted in the references wherever possible. Some obvious typographical errors were found and corrected.

We wish to express our thanks to Dr. B. J. DuWaldt, Dr. P. Swerling, and J. E. Stacey who have read the manuscript in translation and have made many valuable suggestions. Further acknowledgement is due to Mrs. Hanna Stadler for expertly typing the manuscript.

Los Angeles, California
May, 1967

ERNEST JACOBS
K. N. TRIROGOFF

Introduction

This book is devoted to some of the problems encountered in the theory of sophisticated signals used in radar. The term *sophisticated signal* is understood to mean a signal for which the product of the signal duration by the spectrum width substantially exceeds unity. Although it is impossible to draw an exact borderline between simple and sophisticated signals, the term "sophisticated signal" is sufficient to define one of the principal characteristics of modern radar.

Recently, various sophisticated signals (frequency-modulated pulses, coded groups, phase-modulated signals, etc.) have found use in radar. This makes it possible to improve the resolution, to ensure simultaneous measurements of the range and range rate of a target, to elecrically scan over finite angular dimensions, etc. Although the realization of such potentialities is associated with substantial difficulties, one can say with certainty that "classical" radar technology, which uses simple signals at constant frequency and duty cycle, yields to more complex methods based on the use of wide-band signals of the sophisticated structure.

The properties of radar signals, which characterize the measurement of a target's range and range rate, are described by the *Woodward ambiguity function*. The role of this function is similar to that of the antenna pattern, i.e., the ambiguity function defines the accuracy and resolution of the range and range rate measurements to the same extent as the antenna pattern defines the accuracy and resolution of the azimuth and elevation measurements. However, in contrast to the antenna pattern, the ambiguity function possesses the important property of invariance. It was first shown by Woodward that by changing the signal structure, it is possible to vary the accuracy and resolution of the range or range rate measurement but that the combined accuracy and combined resolution (of the range and range rate measurements) do not change. There are uncertainties in both the range and range rate measurements, but they do not depend on the form of the signal.

Here, there is an analogy with the *Heisenberg uncertainty principle* of physics according to which the velocities and coordinates of microparticles cannot be measured with an arbitrary high accuracy. However, it must be remembered that the uncertainty principle of physics is due to the interaction of the microparticles, quantization of energy, and some other factors manifesting themselves only on the microscopic scale. The analogy between radar and quantum physics is of a formal nature.

Essentially, similar limitations are encountered in both radar and radio engineering measurements as in quantum physics. In particular, the limitation of frequency resolution in spectral analysis, associated with a finite observation time, can be interpreted as a type of uncertainty relationship.

Considering the exceptional importance of the uncertainty relationships in radar, it seems appropriate to investigate them in a general sense, including the uncertainty principle of physics and other related problems.

A complete treatment of this phenomenon is rendered by the statistical theory of radar. By defining the error probabilities of the target measurement and resolution parameters, the statistical theory fully reveals the role of the ambiguity function and its invariance. However, in our opinion some "deterministic" approaches possessing even greater clarity can also be used in conjunction with the statistical methods. This applies at least to those problems where (as in the case of frequency measurements) the deterministic treatment is traditional. Such an approach may serve as a useful addition to the rigorous, but rather complicated, statistical theory of radar resolution and measurement.

The first part of the book treats the radar uncertainty principle. This material is more academic than significant. The second part is devoted to a problem of definite practical significance, i.e., the synthesis of signals which realize the desired ambiguity function.

The first type of sophisticated signal used in radar was the frequency-modulated pulse with a linearly changing frequency. Subsequently, other types of signals were proposed: FM pulses with special modulation laws, phase-modulated signals, etc. Naturally, the search for new signal forms was at first semi-intuitive. However, in a large number of recent articles, analytical methods of synthesis are investigated to determine the optimal (in one or another sense) signal forms, improvement of the optimality criteria, etc. Unfortunately, the solution of these problems is associated with great mathematical difficulties, and the general methods for the synthesis of signals by ambiguity functions thus far developed are inadequate.

However, it seems timely to systematize the known synthesis methods and to present them in consecutive order. Some of the methods developed for the synthesis of antenna patterns, for choosing optimal signals in communication systems, etc., can also be used to an advantage for designing radar signals. It may also prove useful to borrow from quantum physics where related mathematical laws are being investigated. An attempt is made in this book to consider similarities between radar and adjacent fields. At the same time, the book emphasizes the difference between the disciplines, and clarifies the specific difficulties in synthesizing signals using ambiguity functions.

The remarks made by M. E. Al'bats, L. M. Rogov, E. L. Feinberg, and A. M. Frenkel', who have read various sections of the manuscript, and also by the reviewers, Ya. D. Shirman and V. P. Yakovlev, were taken into consideration; some of the material used in §36 was kindly submitted by L. E. Varakin. To all these persons, the author expresses his sincere gratitude.

Contents

Chapter 3. The Ambiguity Function in the Statistical Theory of Radar

Chapter 4. Synthesis of Signals Using Ambiguity Functions

Appendices

I

Pulse Compression Signals

[§1] OPTIMAL RECEPTION OF SIGNALS—MATCHED FILTER

The problem of optimal processing of radar signals may be formulated as follows. A signal, $s(t)$, reflected from a target is mixed with noise, $n(t)$, when it enters the input of a receiver:

$$x(t) = s(t) + n(t). \tag{1–1}$$

The input of the receiver, $x(t)$, is transformed into another signal, $y(t)$, which is uniquely related to $x(t)$. The receiver acts on the input function $x(t)$ as a mathematical operator, $L(x)$:

$$y(t) = L[x(t)]. \tag{1–2}$$

The output signal, $y(t)$, is then used for determining the target's parameters, *i.e.*, the range, range rate, angular coordinates, etc.

Measurements become possible only after the presence of a reflected signal has been established, *i.e.*, after the target's detection. It is necessary that the reflected signal level at the receiver output be greater than the noise and other interferences. Therefore, it is assumed that the receiver is optimized to ensure the maximum signal-to-noise ratio. Under very general assumptions, this condition uniquely defines the receiver parameters.

First of all, let it be assumed that the receiver is a linear or quasilinear electrical circuit so that superposition is valid, *i.e.*,

$$L(x) = L(s+n) = L(s) + L(n).$$

Many linear transformations can be represented by the integral form,

$$L(x) \equiv y(t) = \int_{-\infty}^{\infty} x(t')\, h(t, t')\, dt'. \tag{1–3}$$

In most of the important cases encountered in radar, the transformation kernel $h(t, t') = h(t - t')$; hence,

$$y(t) = \int_{-\infty}^{\infty} x(t')\, h(t - t')\, dt'. \tag{1–4}$$

As is well known, the kernel $h(t-t')$ corresponds to any linear passive circuit. However, this form of kernel is even somewhat more general. If time shifts are preserved during transformation, *i.e.*, if the output signal, $y(t-t_0)$, corresponds to an input signal, $x(t-t_0)$, and, moreover, if this holds true for any function, $x(t)$, and for any value of t_0, then such a transformation can also be represented by the integral (1–4) (Khurgin & Yakovlev [71], p. 15). If the above condition is not satisfied, then it means that the receiver characteristics change substantially with time. Although such changes are encountered in some special cases (for example, time control of amplification), it is quite clear that they are not a common property of radar receivers.

As a result of these assumptions, the receiver has been idealized by replacing it with an equivalent, linear frequency-selective passive circuit, *i.e.*, by a filter. The transformation kernel, $h(t)$, may be interpreted as the impulse response of this filter, and the problem of finding the optimal reception method reduces to determining a function, $h(t)$, for which the maximum signal-to-noise ratio is achieved at the filter's output. The impulse response, $h(t)$, is uniquely related to the frequency characteristic (transmission factor) of the filter:

$$h(t) = \frac{1}{2\pi} \int_{-\infty}^{\infty} H(\omega)\, e^{j\omega t}\, d\omega ,$$
$$H(\omega) = \int_{-\infty}^{\infty} h(t)\, e^{-j\omega t}\, dt ; \tag{1–5}$$

therefore, instead of $h(t)$, the optimum frequency characteristic, $H(\omega)$, may be sought.

By virtue of the linearity of the transformation, the signal at the output of the filter consists of two parts: $y_s(t)$ and $y_n(t)$. The former is only a function of the message signal, $s(t)$, while the latter is only a function of the noise, $n(t)$. Denoting $S(\omega)$ as the spectrum of the message signal, $s(t)$, the output component due to the signal is

$$y_s(t) = \frac{1}{2\pi} \int_{-\infty}^{\infty} S(\omega)\, H(\omega)\, e^{j\omega t}\, d\omega . \tag{1–6}$$

The output power of the message signal, except for a constant multiplier, can be represented (as a function of time) by

$$w_s(t) = |y_s(t)|^2 = \frac{1}{4\pi^2} \left| \int_{-\infty}^{\infty} S(\omega)\, H(\omega)\, e^{j\omega t}\, d\omega \right|^2 .$$

It is assumed that the reception is against a background of stationary noise, $n(t)$, the spectrum power density of which is $N(w/Hz)$. The noise power at the output is

$$w_N = \frac{N}{2\pi} \int_{-\infty}^{\infty} |H(\omega)|^2 \, d\omega .$$

Therefore, the *signal-to-noise* ratio (with respect to power) is given by

$$\varrho(t) = \frac{w_s(t)}{w_N} = \frac{\left| \int_{-\infty}^{\infty} S(\omega)\, H(\omega)\, e^{j\omega t}\, d\omega \right|^2}{2\pi N \int_{-\infty}^{\infty} |H(\omega)|^2 \, d\omega} . \tag{1–7}$$

Thus the problem of finding the optimal reception method for a given signal function, $S(\omega)$, reduces to finding a function, $H(\omega)$, for which the function $\varrho(t)$ is a maximum at the time, t. At the time t, the signal-to-noise ratio is maximum; at other instants of time the signal level is lower, but the noise remains the same (since the latter is a stationary random process).

Applying the Bunyakovskii-Schwartz inequality to the numerator of (1–7), we obtain:*

$$\varrho(t) \leqslant \frac{\int_{-\infty}^{\infty} |S(\omega)|^2 \, d\omega \int_{-\infty}^{\infty} |H(\omega)|^2 \, d\omega}{2\pi N \int_{-\infty}^{\infty} |H(\omega)|^2 \, d\omega} = \frac{1}{2\pi N} \int_{-\infty}^{\infty} |S(\omega)|^2 \, d\omega . \tag{1–8}$$

Thus, for any filter characteristic, $H(\omega)$, the signal-to-noise ratio, $\varrho(t)$, cannot exceed the maximum value

$$\varrho_{\max} = \frac{1}{2\pi N} \int_{-\infty}^{\infty} |S(\omega)|^2 \, d\omega = \frac{E}{N}, \tag{1–9}$$

where E = the total input signal energy.**

It will be shown now that the above maximum value of ϱ is attained when

The Bunyakovskii-Schwartz inequality, which will be used frequently in the book is

$$\left| \int f_1(\omega) f_2(\omega)\, d\omega \right|^2 \leqslant \int |f_1(\omega)|^2 \, d\omega \int |f_2(\omega)|^2 \, d\omega .$$

** Here and henceforth, E stands for the energy of a complex signal $s(t)$, which is double the actual energy. For more details, see §18.

the filter frequency characteristic, $H(\omega)$, is chosen from the criterion

$$H(\omega) = \alpha S^*(\omega)\, e^{-j\omega\tau}, \tag{1–10}$$

where α = an arbitrary complex constant,
τ = some quantity (real) having the dimension of time,
$S^*(\omega)$ = the complex conjugate of the signal spectrum.
Substituting (1–10) into (1–7) yields

$$\varrho(t) = \frac{|\alpha^2| \left| \int\limits_{-\infty}^{\infty} |S(\omega)|^2 \, e^{j\omega(t-\tau)} \, d\omega \right|^2}{2\pi |\alpha^2| N \int\limits_{-\infty}^{+\infty} |S(\omega)|^2 \, d\omega}.$$

When $t=\tau$, this becomes

$$\varrho(\tau) = \frac{\left(\int\limits_{-\infty}^{\infty} |S|^2 \, d\omega \right)^2}{2\pi N \int\limits_{-\infty}^{\infty} |S|^2 \, d\omega} = \frac{1}{2\pi N} \int\limits_{-\infty}^{\infty} |S(\omega)|^2 \, d\omega = \frac{E}{N} = \varrho_{\max},$$

which agrees with (1–9).

Thus, the filter frequency characteristic of (1–10) results in a maximum signal-to-noise ratio at time $t=\tau$ and, consequently, is optimum.

Eq. (1–10) establishes the important dependence of receiver parameters on the spectral composition of the incoming signal for optimal reception. Some examples are given below to illustrate this dependence, but first it is desirable to list some general statements concerning optimal receivers.

(1) To improve signal reception against a background of noise, it is necessary to have high receiver amplification at the signal frequency component with the maximum amplitude and to have the amplification decrease with frequency according to the signal spectrum law. This fact is reflected in (1–10), according to which the magnitude of the optimal receiver's frequency characteristic is defined by the relationship

$$|H(\omega)| = |\alpha| \, |S(\omega)|. \tag{1–11}$$

The factor $|\alpha|$ characterizes the receiver amplification, and its value does not affect the signal-to-noise ratio. Therefore, the value of $|\alpha|$ may be arbitrary. The phase of the signal at the receiver output, characterized by the argument of the complex number α, is also arbitrary.

(2) As noted previously, maximum signal-to-noise ratio of an optimal receiver is obtained at the instant of time $t = \tau$. Therefore, the τ in (1–10) characterizes the signal delay in the receiver. The value of this delay is of interest from the viewpoint of physically realizing the filter (see §2). Except for the above aspect, the delay may be arbitrary.

Thus, if the amplification, delay, and the initial phase are not taken into account, the *frequency characteristic of an optimal receiver is the complex conjugate of the received signal spectrum*. Such a receiver is called the *matched filter* receiver. The matching considers both the amplitude and phase structure of the signal spectrum. Consideration of the latter is particularly important for wide-band sophisticated signals.

(3) For optimal reception of a signal against a background of stationary white noise by a matched filter receiver, the *excess* of signal over noise does not depend on the pulse duration, spectrum width, or other characteristics of the pulse. The value of $\varrho_{\max}$ is only a function of *signal energy and noise density*.

(4) It was assumed in the preceding development that the spectrum of the reflected signal is completely known. However, even for the simplest case of a single, point target, the reflected signal contains two unknown parameters – Doppler frequency shift, Ω, and delay time, t_0. In addition, such a signal also has an arbitrary amplitude, A, and an arbitrary initial phase, θ. If the spectrum of the transmitted signal is $S_0(\omega)$, the spectrum of the reflected signal has the form,

$$S(\omega) = A S_0(\omega - \Omega)\, e^{-j(\omega t_0 - \theta)} .$$

Therefore, a receiver with the frequency characteristic,

$$H(\omega) = A\alpha\, e^{-j\theta} S_0^*(\omega - \Omega)\, e^{-j\omega(\tau + t_0)} = \alpha' S_0^*(\omega - \Omega)\, e^{j\omega\tau'} , \qquad (1\text{–}12)$$

(where α' and τ' are new arbitrary quantities) is needed for optimal reception of the reflected signal.

Hence, it is clear that the reflected signal's delay, amplitude, and initial phase do not affect the requirements for optimal reception. If there is no Doppler shift, a receiver matched to the transmitted signal performs optimal signal processing. However, variations in the carrier frequency of the reflected signal require variations in the receiver parameters for optimal processing [the function $S_0^*(\omega)$ is replaced by $S_0^*(\omega - \Omega)$].

Since the Doppler frequency shift is defined by the range rate of the target and is not known beforehand, optimal reception of signals reflected from moving targets cannot be accomplished by only one matched filter. Apparently, maximum signal-to-noise ratios will be obtained using a bank of matched filters with incremented frequencies. The target's range rate (Doppler frequency) can be determined in this case from noting the channel in which the signal-to-noise ratio is maximum, and the target's range from the signal delay in this channel. Such a method of reception will be considered in the subsequent sections.

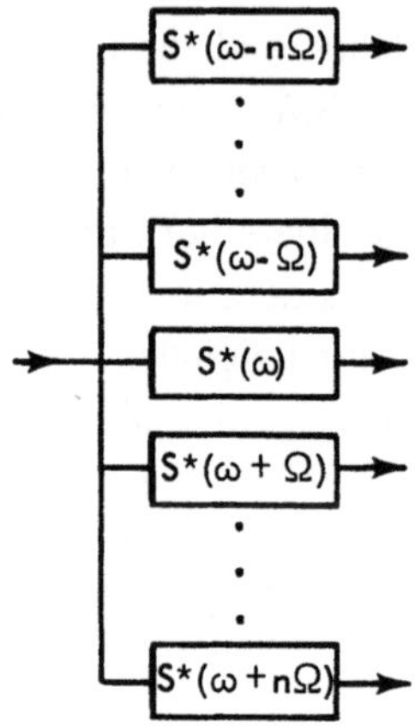

Fig. 1

In conclusion, it is noted that the energy criterion of optimality used above cannot be considered as altogether rigorous. In a more accurate statement of the problem, the condition of greatest probability of obtaining the correct solution in detecting signals and measuring their parameters is chosen as the optimality criterion. Some of these problems are considered in Chapter 3. The statistical treatment provides an additional confirmation of optimal properties of matched filters and the multi-channel circuit shown in Figure 1. It is also noted that the case of reception against a background of white noise considered here, can be readily generalized for a noise background of any arbitrary power spectrum, $N(\omega)$, (see Wainstein & Zubakov [8], p. 80).

[§2] RESPONSE OF A PULSED SIGNAL TO A MATCHED FILTER—CORRELATION METHOD

It has been shown that the frequency characteristic of an optimal receiver is related to the signal spectrum as follows (for the sake of simplicity, the arbitrary constant is taken as unity):

$$H(\omega) = S^*(\omega)\, e^{-j\omega\tau}.$$

Therefore, the output signal of an optimal receiver can be written in the form

$$y_s(t) = \frac{1}{2\pi} \int_{-\infty}^{\infty} S(\omega)\, S^*(\omega)\, e^{j\omega(t-\tau)}\, d\omega\,. \tag{2–1}$$

Furthermore, it follows from the definition of spectral density that

$$\begin{aligned} y_s(t) &= \frac{1}{2\pi}\int d\omega \int S^*(\omega)\, s(t')\, e^{j\omega(t-t'-\tau)}\, dt' \\ &= \frac{1}{2\pi}\int s(t')\, dt' \int S^*(\omega)\, e^{-j\omega(t'-t+\tau)}\, d\omega \\ &= \int_{-\infty}^{\infty} s(t')\, s^*(t'-t+\tau)\, dt'\,. \end{aligned} \tag{2–2}$$

This transformation of (2–1) results in an expression for the output signal, $y_s(t)$, which is in terms of the input signal, $s(t)$, and not in terms of its spectral density. It is noted that such a transformation is equivalent to applying the convolution theorem. Comparing this representation of the output signal with the general expression (1–4), it is seen that for optimal reception the transformation kernel has the form,

$$h(t) = s^*(-t+\tau)\,. \tag{2–3}$$

The kernel, $h(t)$, is the matched filter's impulse response, *i.e.*, the response to the input signal of a short-duration pulse (δ-function). The expression (2–3) is the mirror image of the complex conjugate of the transmitted signal but shifted by the time τ.

The magnitude of the delay is important from the viewpoint of filter realization. It can be readily seen that the delay, τ, must not be smaller than the duration of the transmitted pulse:

$$\tau \geqslant T\,.$$

Otherwise, the response at the filter output would lead the input pulse. Such a condition for filter realization is the only requirement for signals of finite duration (see, for example, Gonorovskii [13], p. 634). However, this condition does not indicate how to synthesize a filter having the impulse response of (2–3). In addition, considering the engineering task, it is easy to note some difficulties in the statement of the synthesis problem.

For lumped parameter circuits, the filter impulse response characteristic cannot be bound over all time and, consequently, matching to a signal of finite duration is not possible. Signals of infinite duration (a signal spectrum having an arbitrarily rapid attenuation beyond the "cutoff" frequency) require, strictly speaking, that the delay in the filter be infinite, which is also not possible.

With the use of distributed parameter circuits (transmission lines), it becomes possible to synthesize accurately filters for most signals of finite duration (see Lezin [43], pp. 41, 56; Shirman & Golikov [75], p. 138). However, the practical potentialities of using lossless transmission lines are not great; often ladder networks or similar devices are used instead of the transmission line. Therefore, while matched filters are almost always synthesized by using an approximate solution, it is desirable to obtain as close an approximation as possible. Theoretical and practical methods for synthesizing matched filters can be found in the special literature (for example, the book by Lezin mentioned previously); in this text it is assumed that filter synthesis for any signal is possible. In the remaining portions of the book, filter realization is not of importance and thus the delay is not considered; therefore, (2–2) can be rewritten in the form

$$y_s(t) = \int_{-\infty}^{\infty} s(t')\, s^*(t' - t)\, dt' . \tag{2–4}$$

Because of the analogy with the theory of random processes, the integral on the right-hand side of (2–4) is called the *autocorrelation function of a complex determined signal* (*i.e.*, a signal of known form).* The signal at the receiver input contains not only the message signal but is a mixture of the message signal and noise:

$$x(t) = s(t) + n(t) .$$

The total signal at the output of the optimal receiver is expressed as follows:

$$y(t) = \int_{-\infty}^{\infty} x(t')\, s^*(t' - t)\, dt' . \tag{2–5}$$

Thus, the optimal receiver computes the *correlation function* between the input signal, $x(t)$, and the expected signal, $s(t)$. The right-hand side of (2–5) is frequently called the *correlation integral.*

The analogy between the output of an optimal receiver and the correlation

* It is noted that this analogy is far from being complete since none of the statistical properties are reflected by this equation. In subsequent sections, this subject is considered in greater detail.

function shows that it is possible to design an optimal receiver not only with a matched filter but also using a correlator that is associated with random processes. A schematic diagram of such a correlator circuit is given in Figure 2. The receiver contains a multiplier with two inputs. The input signal, $x(t')$, is applied at one of the inputs, and a known function $s^*(t')$ is applied to the other input with a time delay of t. The product $x(t')\cdot s^*(t'-t)$ is formed at

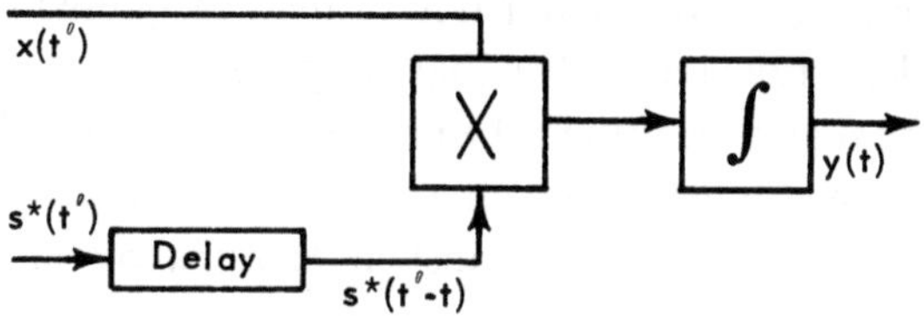

Fig. 2

the output of the multiplier. This product is then fed into the integrator which performs the integration over a sufficiently large period of time. As a result, the integral (2–5) is computed, *i.e.*, $y(t)$ is found as a parameter of t which is defined by the delay line length. By repeating the integration for a different delay, the value of $y(t)$ can be computed for another instant of time. To determine the function $y(t)$ over a given time interval, it is necessary to integrate for several values of t in the interval and to perform the integration over limits which substantially exceed the time interval of the function $y(t)$. To do this it is necessary to record the input signal $x(t)$ and to repeat the calculation for various delay times. An alternative is to use several circuits analogous to that shown in Figure 2 with different delay line lengths. Thus, $y(t)$ is computed for different times at the different outputs.

Therefore, correlators are either multi-channel devices (even when there is only one unknown parameter, *i.e.*, the time of arrival), or do not provide an output response as the input is received. This fact substantially limits the use of correlators in radar.

The advantage of correlators is their flexibility which makes possible a quick transition to various signal forms. To do this, it is only necessary to change the function $s^*(t)$ supplied to the multiplier. It should be kept in mind that the optimal filter is matched to only one signal form, and it would be necessary to change the filter circuit substantially to match to another signal. Regardless of the differences in the two reception methods considered above, the results of the processing (*i.e.*, the form of the output signal) are identical in both cases. Apart from the realization methods, the correlator and the matched filter can be considered as completely equivalent operators, and in the subsequent sections, no distinction will be made between these two methods.

[§3] OPTIMAL RECEPTION OF FREQUENCY-MODULATED PULSE SIGNALS

For signals with comparatively simple envelopes (rectangular, trapezoidal, etc.) and with a constant carrier frequency, optimal matching of the receiver does not produce any new qualitative results. Such matching renders only a slight energy gain when compared to the widely used reception circuits as, for example, a tuned amplifier that is approximately matched to the bandwidth of the signal spectrum. The improvement usually does not exceed 1 *dB* and the matched filter can be interpreted as one requirement for improving the usual type of receiver (Wainstein & Zubakov [8], p. 91; Lezin [43], p. 104).

For a packet of coherent radio pulses, the optimal receiver requires a comb filter (recirculator). This can readily be seen by using the spectrum structure of the sequence of coherent pulses in (1–10). In this case, optimal matching also results in some improvement of the reception.

However, new results are obtained for sophisticated signals, *i.e.*, signals with wide spectra and time bandwidth products substantially exceeding unity. To avoid mathematical difficulties, a special signal of this type is considered in this section. Let it be assumed that the transmitted pulse has a Gaussian (bell-shaped) envelope, and that the instantaneous frequency changes linearly with time, *i.e.*,

$$a(t) = e^{-\frac{t^2}{T^2}},$$
$$\omega_c(t) = \omega_0 + \gamma t.$$

Thus, the sophisticated signal $s(t)$ has the form

$$s(t) = e^{-\frac{t^2}{T^2}} e^{j(\omega_0 t + \gamma t^2)}. \tag{3–1}$$

Here T is one-half the pulse duration between the 0.37 levels of the maximum amplitude.

Since the instantaneous frequency of the signal changes linearly, it can be negative. In this connection, it is assumed that the mean frequency ω_0 is sufficiently high so that

$$|2\gamma|\, T \ll \omega_0,$$

and thus the region of negative frequency corresponds to negligibly small signal amplitudes.

First of all, it is necessary to determine the spectrum of the pulse:

$$S(\omega) = \int_{-\infty}^{\infty} s(t)\, e^{-j\omega t}\, dt = \int_{-\infty}^{\infty} e^{-\left[\left(\frac{1}{T^2} - j\gamma\right) t^2 + j(\omega - \omega_0) t\right]} dt.$$

Introducing the notation

$$1 - j\gamma T^2 = \mu^2, \tag{3-2}$$

the following is obtained after a simple transformation:

$$S(\omega) = \frac{T}{\mu} \mathrm{e}^{-\frac{(\omega-\omega_0)^2 T^2}{4\mu^2}} \int_{-\infty}^{\infty} \mathrm{e}^{-z^2} dz = \frac{T\sqrt{\pi}}{\mu} \mathrm{e}^{-\frac{(\omega-\omega_0)^2 T^2}{4\mu^2}}$$

In accordance with the adopted notation, μ is a complex number. Denoting the square of the peak value of this number by m,

$$m = |\mu^2| = \sqrt{1 + \gamma^2 T^4}, \tag{3-3}$$

and separating the real and the imaginary parts in the exponent,

$$S(\omega) = \frac{2\sqrt{\pi m}}{\Omega} \mathrm{e}^{-\frac{(\omega-\omega_0)^2}{\Omega^2}} \mathrm{e}^{j\left[\gamma T^2 \frac{(\omega-\omega_0)^2}{\Omega^2} - \frac{1}{2}\tan^{-1}\gamma T^2\right]}, \tag{3-4}$$

where

$$\Omega = \frac{2m}{T} = 2\frac{\sqrt{1+\gamma^2 T^4}}{T} \tag{3-5}$$

is the effective spectrum width.

Thus, the amplitude spectrum of a Gaussian pulse with linear frequency modulation is also Gaussian and centered about the mean frequency, ω_0. The higher the frequency rate of change, γ, the wider the spectrum. The dependence of the effective spectrum width on the parameter γT^2 is given by (3–5) and is plotted on Figure 3.

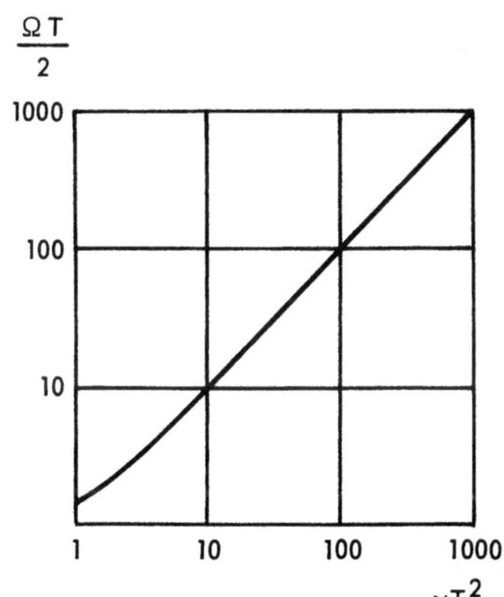

Fig. 3

The essential feature of a frequency-modulated pulse's spectrum is the complicated dependence of phase on frequency. This dependence is given by the last factor in (3–4), and, as will be made clear in the subsequent sections, it is this dependence that defines the specific properties for the optimal processing of signals. To clarify this principle, a simplified method of processing is first considered.

According to the general rule, the frequency characteristic of the matched filter must be the complex conjugate of the signal spectrum (to within a negligibly small amount). From (3–4), this frequency characteristic has the form

$$H(\omega) = e^{-\frac{(\omega-\omega_0)^2}{\Omega^2}} e^{-j\left[\gamma T^2 \frac{(\omega-\omega_0)^2}{\Omega^2} - \frac{1}{2}\tan^{-1}\gamma T^2\right]}. \tag{3–6}$$

The amplitude (magnitude) of the characteristic,

$$|H(\omega)| = e^{-\frac{(\omega-\omega_0)^2}{\Omega^2}},$$

is the same as the amplitude spectrum of the signal, which makes it possible to achieve the best reception of the signal against a background of fluctuating noise. However, the exact form of the amplitude is frequently of no particular importance. First, let it be assumed that the amplitudes of all the spectral frequency components are unchanged when transmitted, and that the filter's phase characteristic is matched to the signal's phase spectrum, *i.e.*,

$$H_\varphi(\omega) = e^{-j\left[\gamma T^2 \frac{(\omega-\omega_0)^2}{\Omega^2} - \frac{1}{2}\tan^{-1}\gamma T^2\right]}. \tag{3–7}$$

If a frequency-modulated (FM) pulse signal is transmitted through such a filter, all the components of the signal spectrum become compressed and, from (3–4), the output is

$$y_s(t) = \frac{1}{2\pi}\int_{-\infty}^{\infty} S(\omega)\,H_\varphi(\omega)\,e^{j\omega t}\,d\omega = \frac{\sqrt{m}}{\sqrt{\pi}\,\Omega}\int_{-\infty}^{+\infty} e^{-\frac{(\omega-\omega_0)^2}{\Omega^2}+j\omega t}\,d\omega$$
$$= \sqrt{m}\, e^{-\frac{t^2}{(T/m)^2}} e^{j\omega_0 t}. \tag{3–8}$$

Thus, a filter with the characteristic of (3–7) eliminates the frequency modulation within the pulse, and as a result, the output is a Gaussian amplitude-modulated signal. Comparing this output signal with the input pulse (3–1), it is seen that the pulse duration is decreased by the factor m, *i.e.*, the pulse duration is compressed by a factor of m and the pulse amplitude increased by $\sqrt{m}$. Such a result can be explained as follows. The initial FM signal has a broad spectrum. The frequency band of this signal is defined mostly by the frequency change within the pulse and, to a much lesser extent, by the duration of the pulse envelope. This spectrum corresponds, generally speaking, to a short-duration pulse. However, the frequency components of the signal have different phases, and the superposition of all the frequency components results in a long-duration pulse. The filter characteristic (3–7) is matched to the signal's phase spectrum. This means that the filter "adjusts" (with respect to phase) the spectrum components in such a manner that the signal is compressed at the output. Therefore, at a particular instant of time (at $t=0$), the amplitudes of spectral components are summed arithmetically and the peak of the signal substantially increases. The filter of (3–7) does not change the amplitudes of the spectral components and, consequently, the signal energies at the input and output are identical. Therefore, the amplitude increase produces a corresponding decrease in the signal duration. From the requirement of energy balance, a decrease in the pulse duration by a factor of m is accompanied by an increase in the amplitude by a factor of $\sqrt{m}$; this is expressed in (3–8). By putting the spectral components in phase, the filter compresses the pulse to a duration which corresponds to the spectrum width; moreover, pulse compression occurs without loss of signal energy.

For the sake of simplicity, amplitude changes of the signal frequency components were not considered. The filter with the characteristic of (3–7) transmits a wide spectrum of noise without any constraints and, consequently, is unsuitable for the reception of a signal against a background of broadband noise. The optimal filter must be matched to the signal spectrum not only in phase but also in amplitude, *i.e.*, its characteristic must be defined by (3–6). Such a filter characteristic can be realized by cascading two filters (see Fig. 4). The first filter has the characteristic

$$H_\alpha(\omega) = e^{-\frac{(\omega-\omega_0)^2}{\Omega^2}}$$

and ensures amplitude constraints in accordance with the amplitude spectrum. The second filter has the characteristic $H_\varphi(\omega)$ which has been considered previously.

The introduction of additional amplitude filtering does not produce any substantial changes in the pulse compression process. In fact, considering

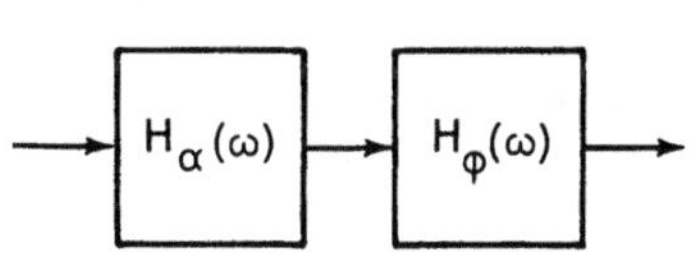

Fig. 4

(3–4) and (3–6), the output signal is

$$y_s(t) = \frac{\sqrt{m}}{\sqrt{\pi}\,\Omega} \int_{-\infty}^{+\infty} e^{-\frac{2(\omega-\omega_0)^2}{\Omega^2} + j\omega t}\, d\omega = \sqrt{m'}\, e^{-\frac{t^2}{(T/m')^2}}\, e^{j\omega_0 t}, \qquad (3\text{–}9)$$

where

$$m' = \frac{m}{\sqrt{2}} = \sqrt{\frac{1+\gamma^2 T^4}{2}}. \qquad (3\text{–}10)$$

Consequently, the addition of the amplitude restriction leads to a decrease in the compression ratio by a factor of only $\sqrt{2}$. This can be explained by the fact that the signal spectrum becomes somewhat narrower at the output of the filter $H_\alpha(\omega)$, and as a result, an increase in the compressed pulse duration occurs.

Pulse compression makes it possible to improve the accuracy of radar measurements because it reduces the effective duration of the transmitted pulse. It is well known that range accuracy and resolution are usually defined by the duration of the transmitted pulse. Long-duration signals reflected from near targets blend together which lowers the resolution. At best, the target's range is usually measured to within a portion of the pulse duration (the duration is converted into a linear dimension by the expression $\delta_r = Tc/2$, where c is the velocity of light). Therefore, in striving for higher resolution and accuracy, short-duration signals (of the order of one microsecond or less) were used in radars up until recently. On the other hand, the maximum range of a radar is a function of the signal-to-noise ratio and, consequently, as was shown in §1, depends on the energy in the signal pulse.*

The only method for increasing the limited range is to increase the peak power. At present, the peak power of radar transmitters, depending on the wavelength, is hundreds, thousands, and tens of thousands of kilowatts. By the same token, a further increase in peak power is hindered by the power handling limitations of the transmitting components. Therefore, even until recently, the requirements of long range with high accuracy and resolution were considered contradictory, and compromises were sought between these requirements.

The use of long-duration transmitted pulses (the order of tens or hundreds of microseconds) makes it possible to obtain high signal energy, and thus a greater range is possible from the available peak powers. By the same token,

* Usually, the ratio of the transmitter peak power to the receiver sensitivity enters into the range equation. The sensitivity of a receiver is proportional to the spectrum width and, consequently, for simple signals, it is inversely proportional to the pulse duration. Therefore, the range equation contains essentially the product of the peak power by the duration, *i.e.*, the energy of the transmitted pulse.

the use of FM results in a broad spectrum signal and, consequently, in a short-duration pulse after compression. Therefore, high accuracy and resolution are attained for long-range radars. It is clear from the above that the range accuracy and resolution depend not on the duration of the transmitted pulse but on the spectrum width. For simple signals these parameters are uniquely related. The sophisticated signals, in particular FM signals, have a wide spectrum for long durations, and substantial shortening can be attained by optimal processing of the signals.

In the previous development, the following expression was obtained for the compression ratio:

$$m' = \sqrt{\frac{1 + \gamma^2 T^4}{2}}.$$

In order to obtain a larger value of m', it is necessary to increase the parameter $|\gamma|\, T^2$, and in this case

$$m' \approx \frac{|\gamma|\, T^2}{\sqrt{2}}.$$

The following is a clarification of this relationship. As stated previously, the quantity T is one-half the pulse duration between the 0.37 levels of the maximum amplitude. Let the duration (as is frequently done) be measured between the 0.5 levels of the maximum amplitude; then

$$T_{0.5} = 1.66T.$$

The frequency deviation during this time interval is

$$f_d = \frac{\omega_d}{2\pi} = \frac{2\,|\gamma|\, T_{0.5}}{2\pi} = 0.53\,|\gamma|\, T.$$

Considering the above,

$$m' \approx 0.8 f_d T_{0.5}.$$

Thus, with regard to the order of magnitude, the compression ratio is equal to the product of the frequency deviation by the pulse duration. It will be seen that this ratio is the same for many cases.

[§4] COMPRESSION RATIO

Pulse compression is not exclusively a characteristic of the particular signal form considered in the previous section. In all cases where the transmitted signal is substantially widened by modulation, the compression of the pulse can be accomplished during reception.

If $S(\omega)$ is the spectrum of the input signal, the signal $s(t)$ and its autocorrelation function $y_s(t)$ (*i.e.*, the signal at the output of an optimal receiver) are determined from the relationships,

$$s(t) = \frac{1}{2\pi} \int_{-\infty}^{\infty} S(\omega)\, e^{j\omega t}\, d\omega\,, \tag{4–1}$$

$$y_s(t) = \frac{1}{2\pi} \int_{-\infty}^{\infty} |S(\omega)|^2\, e^{j\omega t}\, d\omega\,. \tag{4–2}$$

The qualitative distinction between these expressions lies in that the spectral density of (4–1),

$$S(\omega) = \sigma(\omega)\, e^{-j\psi(\omega)}$$

is a complex function, and that the phases, $\psi(\omega)$, of the separate spectral components depend on frequency in a very complicated manner. In (4–2), the complex quantity $S(\omega)$ is replaced by the positive, real quantity

$$|S(\omega)|^2 = \sigma^2(\omega)$$

with the phases of all the components equal to zero. Therefore, at time $t=0$, all the components in (4–2) are added arithmetically, and a relatively large amplitude is obtained:

$$y_m = y_s(0) = \frac{1}{2\pi} \int_{-\infty}^{\infty} \sigma^2(\omega)\, d\omega\,. \tag{4–3}$$

Without loss generality, the beginning of the time reference can be chosen so that the maximum value of the signal $s(t)$ is also attained at $t=0$, *i.e.*,

$$a_m = |s(0)| = \frac{1}{2\pi} \left| \int_{-\infty}^{\infty} \sigma(\omega)\, e^{-j\psi(\omega)}\, d\omega \right|.$$

Here the spectral components are added vectorially [due to the phase term $e^{-j\psi(\omega)}$] and compensation between the separate components takes place. As a result, the amplitude is relatively smaller. Due to the conservation of energy, this is accompanied by a corresponding change in the pulse duration.

In order to evaluate the degree of compression, it is necessary to agree on how to define the duration of signals and the width of their spectra. In the case of sophisticated signals, various approaches are possible, but for the majority of pulse (or spectrum) forms encountered in practice, the difference is not very great; the approach is usually chosen for calculation convenience.

Assuming that the pulse has the energy E and a maximum amplitude a_m, the duration T is defined by

$$a_m^2 T = E .$$

Similarly, the spectrum width F is defined by

$$\sigma_m^2 F = E ,$$

where σ_m is the maximum value of the spectral density.*

Furthermore, the signal energies at the input and output of the receiver, from (4–1) and (4–2), are given by

$$E_s = \frac{1}{2\pi} \int_{-\infty}^{\infty} |S(\omega)|^2 \, d\omega = \frac{1}{2\pi} \int_{-\infty}^{\infty} \sigma^2(\omega) \, d\omega = y_m ,$$

$$E_y = \frac{1}{2\pi} \int_{-\infty}^{\infty} |S(\omega)|^4 \, d\omega = \frac{1}{2\pi} \int_{-\infty}^{\infty} \sigma^4(\omega) \, d\omega ,$$

and the following is the expression for the compression ratio

$$m = \frac{T_s}{T_y} = \frac{T_s E_s^2}{E_y} = \eta T_s F_s . \tag{4–4}$$

Here

$$\eta = \frac{\int_{-\infty}^{\infty} \sigma_0^2(\omega) \, d\omega}{\int_{-\infty}^{\infty} \sigma_0^4(\omega) \, d\omega} \tag{4–5}$$

and $\sigma_0(\omega)$ is the normalized spectral density:

$$\sigma_0(\omega) = \frac{\sigma(\omega)}{\sigma_m} .$$

Since $\sigma_0(\omega) \leqslant 1$, then $\eta \geqslant 1$. If the signal spectrum is rectangular, then $\eta = 1$; for a triangular spectrum $\eta = \frac{5}{3} = 1.67$; for a "Gaussian" spectrum (such a signal was considered in the previous section), $\eta = \pi / \sqrt{2} = 2.22$. This order of magnitude of η is the same for other signal types provided the spectral density is a comparatively smooth function. Such signals are of prime interest in pulse compression systems and this will be shown in the following sections. Eq. (4–4) shows that the *compression ratio (order of magnitude) equals the product of the signal duration by the spectrum width.*

In "classical" radar technology, rectangular, Gaussian, or other forms of

* Other definitions of duration and spectrum width will be considered in §13.

pulse envelopes are used; the spectrum widths for these pulse shapes are the order of $1/T$. For such signals, the compression ratio is close to unity, and optimal processing does not result in any noticeable pulse shortening. To obtain substantial shortening, different and more complicated signals must be used in which, in addition to amplitude (envelope shaping), other forms of modulation (frequency, phase, pulse-coded, etc.) are also present. Such additional modulation makes it possible to widen the spectrum substantially, to obtain a large compression ratio and, consequently, to increase considerably the resolution of the radar and the accuracy of the range measurement.

[§5] SPECTRUM COMPRESSION

Signal compression has been analyzed with respect to the signal's time duration. It follows from the symmetry of the direct and inverse Fourier transforms that it is also possible to realize symmetry with respect to the spectrum width. In fact, if the signal $s(t)$ is replaced by its magnitude $|s(t)|$ [or by its square $|s(t)|^2$] in the expression

$$S(\omega) = \int_{-\infty}^{\infty} s(t)\, e^{-j\omega t}\, dt,$$

the value of the spectral density increases substantially at $\omega = 0$. This requires that the signal be not only amplitude modulated but also frequency (or phase) modulated as expressed by the complex function

$$s(t) = a(t)\, e^{j\varphi(t)};$$

moreover, the phase factor substantially widens the spectrum $S(\omega)$. As discussed before, the increase in the value of $S(0)$ results in the decrease in spectrum width (due to the conservation of energy).

The basic relationships characterizing such a compression can be derived similar to the previous development by interchanging frequency with time and replacing the spectral function $\sigma(\omega)$ by the amplitude envelope $a(t)$. For instance, the compression ratio given by (4–4) is also expressed by

$$\eta = \frac{\int_{-\infty}^{\infty} a_0^2(t)\, dt}{\int_{-\infty}^{\infty} a_0^4(t)\, dt},$$

where $a_0(t)$ is the normalized envelope.

The spectrum compression results from the phase (or frequency) modulation being removed in the reception process. In this case the output signal has the same envelope (or its square) as the input signal and has a narrow spectrum which corresponds to a pulse of comparatively long duration. For example, the spectrum compression can be accomplished by modulating the frequency of the receiver's local oscillator in such a manner that it follows the frequency of the incoming signal. Thus, a single-frequency signal is fed to the intermediate-frequency amplifier (IF A) and the IF A can be a narrow-band amplifier which has a lower noise level at the output.

In practice, it is more difficult to obtain spectrum compression than compression with respect to time duration. A matched filter optimally processes a signal with any arbitrary time shift. If there is a frequency shift in the signal (*e.g.*, due to target motion), compression is possible only when the frequency modulation within the pulse is linear. However, even if the frequency modulation is linear, it is also necessary to know with sufficient accuracy the signal's time of arrival; otherwise, the received signal may be beyond the passband of the IF A. To eliminate such a deficiency, multi-channel circuits, as well as saw-tooth modulation of the local oscillator, are used.*

It is noted that, contrary to time compression, spectrum compression can only be realized with nonlinear devices. It was shown in §2 that an optimal receiver performing time compression has the following properties. Let $h(t)$ be the response of the optimal receiver to a δ-function input; this response function is related to the input signal by

$$h(t) = s^*(-t).$$

Similarly, for spectrum compression, it is required that $H(\omega)$, the receiver response to the "frequency" δ-function (*i.e.*, a monochromatic signal), be of the following form:

$$H(\omega) = S^*(-\omega).$$

Obviously, the transformation of the signal frequencies into a complex spectrum by a linear passive device is not possible.

Spectrum compression increases the range-rate measurement accuracy and resolution to the same extent as time compression increases the range measurement accuracy and resolution. The ultimate accuracy that can be obtained with time compression is defined only by the length of the pulse

* One of the methods used in communication systems for receiving an FM signal is similar to spectrum compression. This is the so-called tracking reception of FM signals (see, for example, Kartianu [31], p. 552). However, the noise level of communication equipment is very low, which makes it possible to use feedback between the discriminator and the local oscillator. Apparently such a method is of no interest in radar.

envelope. This property will be confirmed subsequently using more general considerations (§17).

It is not difficult to note that the generalized signal compression concept can be extended. For time compression, the spectrum width is preserved, but the pulse duration changes. Conversely, for spectrum compression the pulse duration is preserved, but the spectrum width is changed. It is also possible to change both parameters simultaneously as shown in the following example. Assume that the frequency of the receiver's local oscillator is varied at a slower rate than the signal frequency and, consequently, there is a smaller frequency deviation for the pulse duration. Thus, frequency modulation is preserved at the intermediate frequency but with a smaller spectrum width. Subsequently, the IF signal is processed through a matched filter. The signal time duration is compressed, but the resulting pulse duration at the output of the filter is obviously greater than would be obtained by compressing the initial pulse. Thus, both the pulse spectrum and duration are compressed. It is not difficult to show that the product of TF is the order of unity and it is possible, within certain limits, to vary either T or F while preserving the magnitude of their product. The total compression ratio (with respect to both time duration and spectrum width) is the order of $T_s F_s$, *i.e.*, the same order of magnitude as either the compression for time duration alone or for spectrum alone.

It should be noted that similar transformations of signal time duration and frequency spectrum are also analyzed from another viewpoint in information theory. These transformations are associated with the so-called Gabor information cells and the various versions of compression correspond to the various configurations of these cells (Brillouin [5], p. 99).

[§6] ADMISSIBLE ERRORS IN THE FREQUENCY- (PHASE-) MODULATION LAW FOR SYSTEMS WITH PULSE COMPRESSION

Effective signal compression is possible only for the case where the optimal filter of the receiver and the arriving signal are matched to each other. In this section the requirements for mutually matching the filter to a wide-band FM signal are analyzed. Let us assume that the filter is matched to an "unperturbed" signal, $s_0(t)$, so that

$$H(\omega) = S_0^*(\omega), \tag{6–1}$$

and, moreover, that the frequency characteristic of the signal is distorted, and thus a "perturbed" signal arrives at the receiver input. This signal has the form,

$$s(t) = s_0(t)\, e^{j\varepsilon(t)} = a(t)\, e^{j[\varphi_0(t) + \varepsilon(t)]}, \tag{6–2}$$

where $\varphi_0(t)$ is the phase-modulation function and $\varepsilon(t)$ is its distortion.

An approximate method for calculating the spectra of FM signals is given in Appendix 1 (assuming certain conditions have been satisfied). Using this method, the spectrum of the signal (6–2) is

$$S(\omega) = \sqrt{\frac{2\pi}{|\varphi_0''(t_0)|}}\, a(t_0)\, e^{j\left[\varphi_0(t_0) - \omega t_0 \pm \frac{\pi}{4}\right]} e^{j\varepsilon(t_0)} = S_0(\omega)\, e^{j\varepsilon(t_0)}, \tag{6–3}$$

where t_0 is the time when the instantaneous frequency of the unperturbed signal $s_0(t)$ equals ω.

This time is determined from

$$\omega_c(t_0) = \varphi_0'(t_0) = \omega. \tag{6–4}$$

It follows from (6–3) that the phase errors of an FM signal at any instant of time result directly in errors in the signal spectrum at that time.*

Using (6–1) and (6–3), the following expression is obtained for the output signal of the receiver:

$$y_s(t) = \frac{1}{2\pi}\int_{-\infty}^{\infty} S(\omega)\, H(\omega)\, e^{j\omega t}\, d\omega$$

$$= \frac{1}{2\pi}\int_{-\infty}^{\infty} |S(\omega)|^2\, e^{j[\varepsilon(t_0) + \omega t]}\, d\omega. \tag{6–5}$$

Moreover, according to (6–4), t_0 depends on ω and $\varepsilon(t_0) = \varepsilon(\omega)$. This relationship is particularly simple for the case of linear FM, which is important to this discussion:

$$\omega = \omega_0 + \gamma t_0 \quad \text{or} \quad t_0 = \frac{\omega - \omega_0}{\gamma}.$$

It is necessary to analyze the effect of phase errors on the compressed pulse form. In the general case, *i.e.*, for a signal with an arbitrary spectrum, $S(\omega)$, and an arbitrary error function, $\varepsilon(\omega)$, this problem is extremely complicated. Various signal forms and types of random errors result in various compressed signals which thus become random functions of time. However,

* Eq. (6–3) is obtained under the assumption that the factor $e^{j\varepsilon(t)}$ is slowly varying. It follows from (A1–5) in Appendix 1 that this condition is satisfied if $[\varepsilon'(t)]^2 \ll \omega'_c(t)$, where $\omega_c(t)$ is the rate of change of unperturbed signal's instantaneous frequency, and $\varepsilon'(t)$ is the frequency error. Therefore, (6–3) is only applicable for the case of small frequency distortion.

an approximation of the admissible phase errors is sufficient for the purpose of this discussion and the development can be confined to some representative examples. First, the case of the phase error, $\varepsilon(\omega)$, following the sinusoidal law is considered:

$$\varepsilon(\omega) = \varepsilon_m \sin(\omega T_\varepsilon + \beta).$$

Obviously,

$$e^{j\varepsilon(\omega)} = \sum_{k=-\infty}^{\infty} J_k(\varepsilon_m)\, e^{jk(\omega T_\varepsilon + \beta)},$$

where J_k is the Bessel function.

Substituting this expansion into (6–5),

$$y_s(t) = \sum_{k=-\infty}^{\infty} J_k(\varepsilon_m)\, e^{jk\beta} \frac{1}{2\pi} \int_{-\infty}^{\infty} |S(\omega)|^2\, e^{j\omega(t-kT_\varepsilon)}\, d\omega.$$

The last integral is the autocorrelation function of the undistorted signal, $R(t)$, shifted in time:

$$R(t) = \frac{1}{2\pi} \int_{-\infty}^{\infty} |S(\omega)|^2\, e^{j\omega t}\, d\omega.$$

Finally,

$$\begin{aligned} y_s(t) = {} & J_0(\varepsilon_m)\, R(t) + J_1(\varepsilon_m)\, \{R(t - T_\varepsilon)\, e^{-j\beta} - R(t + T_\varepsilon)\, e^{j\beta}\} \\ & + J_2(\varepsilon_m)\, \{R(t - 2T_\varepsilon)\, e^{-j2\beta} + R(t + 2T_\varepsilon)\, e^{j2\beta}\} + \cdots \end{aligned}$$

If there are no phase distortions ($\varepsilon_m = 0$), the compressed pulse is the autocorrelation function $R(t)$. Due to the parasitic phase modulation of the spectrum, the pulse is modified by the factor $J_0(\varepsilon_m)$ and, in addition, there also appear the "side" pulses decreasing in magnitude. The side pulses are similar to the principal pulse and are separated from each other by the time interval T_ε (see Fig. 5). Such a change in the signal form can substantially lower the gain obtained by compression. The peak of the signal is lower, which effectively is a relative increase in the noise level; the signal energy is distributed

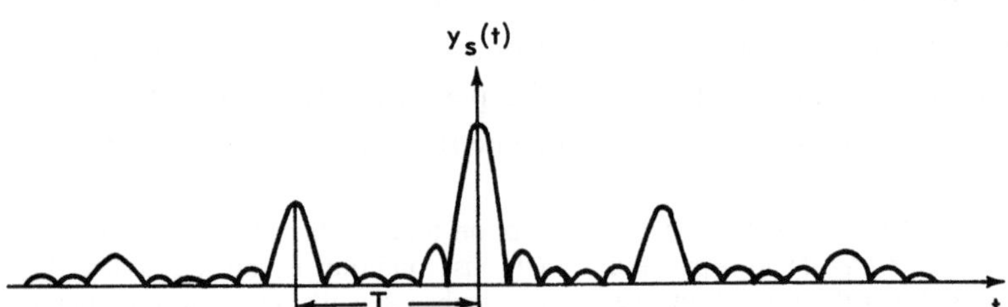

Fig. 5

over a larger time interval and, as a consequence, the range accuracy and resolution are lower; the additional spikes mask the return from nearby targets. If T_ε is small, the side pulses draw together and "blend" into the principal pulse. The latter becomes substantially distorted; its duration increases which, roughly speaking, is equivalent to a decrease in the compression ratio.

The quantitative effect of phase distortion can be estimated from Figure 6. Figure 6a shows the amplitude decrease of the principal pulse, while Figure 6b shows the level of the first two side pulses relative to the principal one. The values in these plots are expressed in *dB*. The experimental data obtained by Cook [37] are shown by small circles on the graph. These results were obtained from the oscillograms of compressed pulses with sine wave phase modulation distortion in the FM signals.

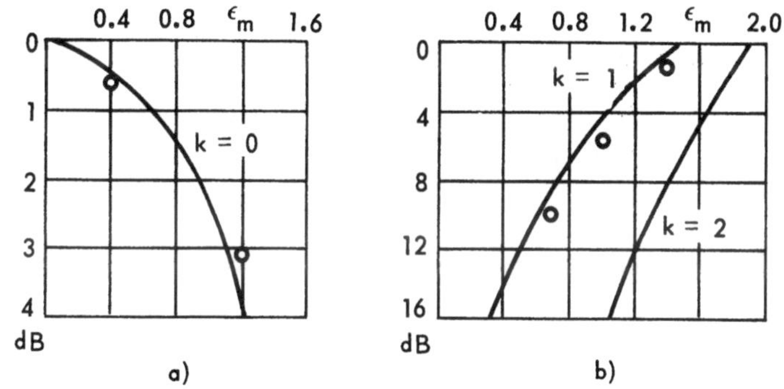

Fig. 6

These graphs show that the form of the compressed pulse deteriorates substantially for phase errors the order of 1–2 *rad*. For such values of ε_m, the level of the principal pulse drops by more than a factor of 2 and, in addition, the sidelobes attain a relatively large magnitude (increase the order of several *dB*). Thus, the admissible level of phase distortion is rather low. It is important for the subsequent discussion that the admissible phase distortion not be dependent on the compression ratio. This assumption is even implied in the derivation of (6–5) which contains the amplitude spectrum, $|S(\omega)|$, but does not account for the phase structure that defines the degree of compression.

Sinusoidal distortions are a rather special case. The only value of this example is to illustrate a simple method for determining the admissible phase errors of an arbitrary form of pulse compression with an arbitrary compression ratio. In actuality, the phase errors are irregular. The phase error magnitude is characterized by the mean square deviation of phase, σ_ψ, and not by the maximum error, ε_m. In addition, the phase deviation function of the frequencies close to the carrier must also be considered. It is natural to assume that the phases of these close frequencies differ negligibly from each other and that the phase errors may be considered as independent only for frequencies that are greatly separated from the carrier. The greater the frequency correlation radius and the smaller the phase change between the spectral components, the smaller the distortion of the spectral pulse. For an identical phase change of all components, the envelope of the compressed pulse does not change. Actually, the distortions depend on the mean square differences of the phases, which are determined for pairs of arbitrary spectrum frequencies separated by the interval $\Delta\omega$. The quantity $\sigma_\psi(\Delta\omega)$ characterizes both the mean square level of the errors and the correlation function.

By considering the statistical nature of phase errors, Green and Moller [18]* estimated their effect. It is assumed in this development that the spectrum envelope is rectangular:

$$|S(\omega)| = \begin{cases} 1 \text{ for } -\frac{\Omega}{2} \leqslant \omega \leqslant \frac{\Omega}{2}, \\ 0 \text{ for all other values of } \omega. \end{cases}$$

Thus, in the absence of distortion, the compressed pulse can be represented by the form

$$y_s(t) = \frac{1}{2\pi} \int_{-\frac{\Omega}{2}}^{\frac{\Omega}{2}} e^{j\omega t}\, d\omega = \frac{\sin \frac{\Omega}{2} t}{\pi t}.$$

It follows from Appendix 1 that such a compressed pulse form corresponds (from the approximation using the stationary phase method) to a rectangular pulse with linear FM. In practice this is a very important case.

In the presence of phase errors, the compressed pulse can be written as the integral

$$y_s(t) = \frac{1}{2\pi} \int_{-\frac{\Omega}{2}}^{\frac{\Omega}{2}} e^{j[\varepsilon(\omega) + \omega t]}\, d\omega, \tag{6–6}$$

where $\varepsilon(\omega)$ is a random function having a normal distribution.

* A similar problem concerning antennas is considered in this paper, but the results are fully applicable to the radar analysis. The analogy is discussed in more detail in Chapter 4.

This distribution can be characterized by the mean square difference of phase, $\sigma_\psi(\Omega)$, for the boundary frequencies of the spectral band, and by the correlation function, $\varrho(\omega)$, which is a measure of the degree of interrelation between the random error of the frequencies close to the carrier. For the purpose of calculation, the correlation function is assumed to be exponential,

$$\varrho(\omega) = e^{-\xi \frac{2\omega}{\Omega}},$$

where ξ is a constant defined as the correlation radius.

As can be seen from Figure 7, various degrees of correlation (from almost "harmonious" [dependent] phase fluctuations [$\xi = 0.286$] to almost independent fluctuations [$\xi = 4$]) depend on the selected value of ξ.

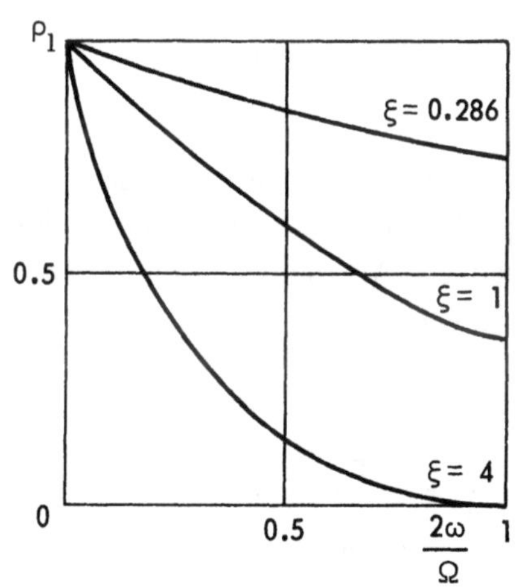

Fig. 7

Further calculations can be made by the statistical trial and error method (the Monte Carlo method). In essence this method reduces to the following. A model of the random function $\varepsilon(\omega)$ having a mean square value of $\sigma_\psi(\Omega)$ and following a given correlation law is chosen. Green and Moller have assumed the thermal noise function for obtaining such models. The mean square value is the effective noise voltage, and the correlation function (with respect to time) is uniquely defined by the frequency selective circuit which filters the noise. For an ordinary RC-integration circuit, the correlation function is exponential. Thus, measured noise oscillograms over one pulse period can be used to simulate (after matching the scales) random models of phase errors with the necessary statistical properties.

Each model of the phase error, $\varepsilon(\omega)$, in its coded form is fed into a computer which calculates the integral (6–6) and consequently determines the random form of the compressed pulse. Figure 8 shows some characteristic pulse forms for several values of the parameters. The undistorted pulse forms (dotted line) are also shown in this figure. By statistically processing (also on a computer) a large number of such calculations, the means of the

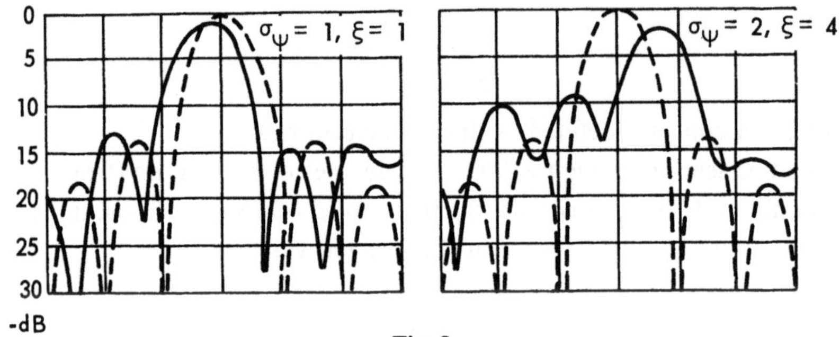

Fig. 8

various parameters of compressed pulses are determined. The computed results of signal amplitude and level of the highest sidelobe are plotted in Figures 9 and 10, respectively. Also shown in Figure 9 are the 95% confidence intervals (vertical lines). The following conclusions which are important to

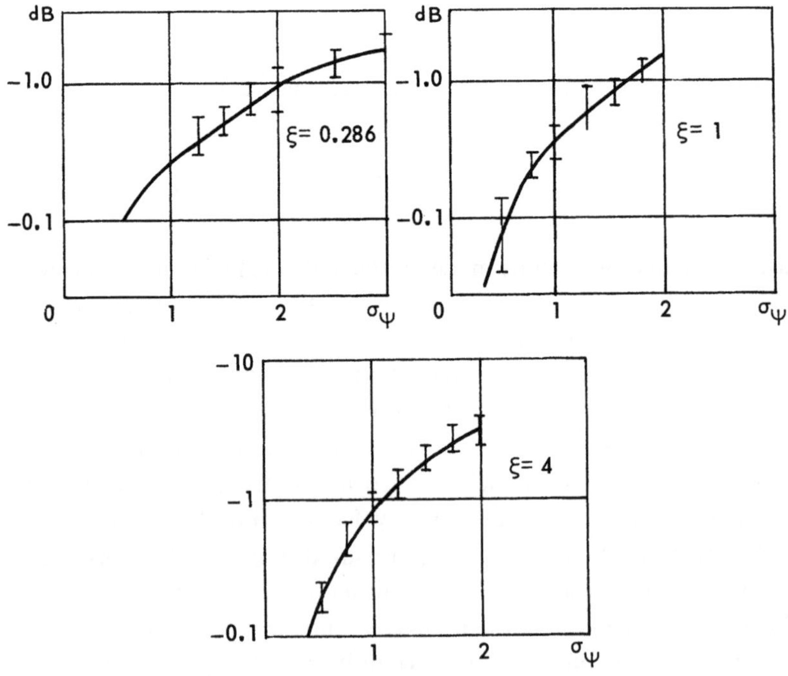

Fig. 9

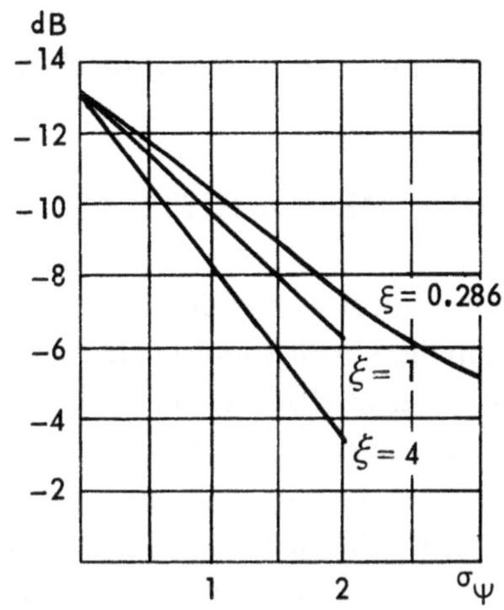

Fig. 10

the subsequent discussion can be drawn from these results (which have been verified experimentally): random phase errors whose mean square value $\sigma_\psi(\Omega)$ is the order of 1–3 *rad* produce substantial deterioration to the compressed signal; the pulse amplitude decreases by 1–2 *dB* (25–40% in power) and the sidelobes increase by 5–10 *dB* (*i.e.*, by a factor of 3–10). Therefore, *it is necessary that the value of $\sigma_\psi(\Omega)$ not exceed several radians to ensure satisfactory compression by the circuit.* Mathematically, this condition is

$$\sigma_\psi \lesssim \lambda\pi\,, \tag{6–7}$$

where λ is a constant the order of unity.

The value of λ is, generally speaking, a function of the radar, signal characteristics, statistical properties of the phase errors (correlation radius), and other factors. However, considering the possible variations in the value of λ, various practical cases are included by condition (6–7) which makes it effectively universal. This estimate of admissible distortion is also applicable to sinusoidal phase variations (see Fig. 6), and is in agreement with the results obtained in other works [6, 19, 38, 76].

It is noted that the total phase change for the pulse duration is the order of $\Delta\Phi \approx T\Omega = 2\pi m$. Therefore, the admissible *relative phase error* estimated by

$$\frac{\lambda\pi}{2\pi m} = \frac{\lambda}{2m}$$

is very small for large values of the compression ratio m. The same estimate is also valid for the relative frequency errors:

$$\frac{\delta\omega}{\Omega} = \frac{\delta\omega T}{\Omega T} \approx \frac{\lambda\pi}{2\pi m} = \frac{\lambda}{2m}\,.$$

These conditions show that for satisfactory pulse compression it is necessary to faithfully follow the FM law defined by the matched filter.

[§7] FORMATION OF SOPHISTICATED SIGNALS

The necessity to follow rigorously the derived modulation law requires the use of multistage amplifiers with sophisticated signals. Frequently the signal is formed at a low-power level at the intermediate frequency, provided the power amplifiers and frequency converters do not introduce large distortions. With such a system it is possible, in principle, to form (at the low-power level) a signal with an arbitrary phase structure and thus accurately match the optimal filter. Here the development is confined to considering only one of the signal forming circuits.

To obtain the needed signal $s(t)$, it is possible to use a linear circuit, *i.e.*, a filter whose frequency characteristic coincides with the signal spectrum:

$$H(\omega) = S(\omega).$$

Supplying to the filter input a pulse of sufficiently short duration (δ-function), the pulse response formed at the filter output, $h(t)$, will be a signal of the required form:

$$h(t) = s(t).$$

Thus, the signal formation problem reduces to the synthesis of a linear filter with the following frequency characteristic:

$$H(\omega) = S(\omega) = \sigma(\omega)\, e^{j\psi(\omega)}. \tag{7–1}$$

Consider a parallel circuit in which each branch contains a narrow-band filter, an attenuator, and a delay line (Fig. 11).

If a short-duration pulse is fed to the system input, each of the filters isolates

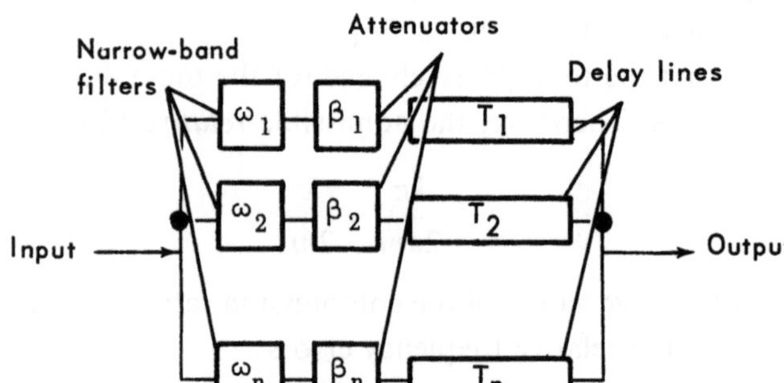

Fig. 11

from this pulse spectrum a narrow band of frequencies corresponding to its pass band.

Assuming that the filters have an ideal rectangular characteristic with a bandwidth Δ* the signal

$$s_k(t) = \frac{1}{2\pi} \int_{\omega_k - \frac{\Delta}{2}}^{\omega_k + \frac{\Delta}{2}} e^{j\omega t}\, d\omega = \frac{\sin \frac{\Delta}{2} t}{\pi t}\, e^{j\omega_k t}.$$

is formed at the output of each filter. Furthermore, this signal is attenuated by a factor of β_k and delayed by the time T_k. As a result, the following composite signal is obtained at the output:

$$s(t) = \sum_{k=1}^{n} \beta_k s_k(t - T_k) = \frac{1}{\pi} \sum_{k=1}^{n} \beta_k \frac{\sin \frac{\Delta}{2}(t - T_k)}{t - T_k}\, e^{j\omega_k (t - T_k)}. \quad (7\text{–}2)$$

This result is the Kotel'nikov series which can be used (with the appropriate choice of the parameters β_k, T_k, ω_k, and the number of terms n) to form in a finite time interval any signal having a finite frequency band (see, for example, Khurgin & Yakovlev [7], p. 107). Thus, assuming that the required signal has a finite spectrum width F, and is of interest only during a finite time interval T (the pulse duration), the conclusion is that any signal form can be produced by the circuit shown in Figure 11.

Signals with a limited spectrum width F have a remarkable property; namely, in order to reproduce exactly such a signal in the time interval T, it suffices to form the signal at $2TF$ discrete points. It is also possible to reproduce the signal by transmitting $2FT$ independent data points of other parameters which are uniquely related to the form of the signal. Therefore, the number $2FT$ is called the *number of degrees of freedom* of a signal. For the circuit shown in Figure 11, such parameters are the values of the amplitudes and phases of the separate spectral components, ω_k, transmitted by the filters.

To develop the basic relationships for the synthesis of the circuit (*i.e.*, to select the proper values for β_k and T_k), it is necessary to consider in somewhat greater detail the transmission of a signal through the delay line. Any physical delay line has a dispersion, *i.e.*, the delay time is a function of the signal frequency. The dispersion of the line is usually characterized by the propagation constant $\gamma(\omega)$ which is also a function of frequency. If the length of the line is l, the frequency ω acquires, during propagation, a phase shift $l\gamma(\omega)$. For a sophisticated signal input, the output voltage is calculated from

* This assumption is not of fundamental significance. Analogous results are obtained for any narrow-band filter characteristic.

the Fourier integral:

$$u_{\text{out}}(t) = \frac{1}{2\pi} \int_{-\infty}^{\infty} S_{\text{in}}(\omega)\, e^{j[\omega t - l\gamma(\omega)]}\, d\omega .$$

The input components, $s_k(t)$, to each delay line in the above circuit cover a narrow frequency band, Δ, about the mean frequency ω_k. It can be shown that under this condition each component propagates along its delay line without noticeable distortion, and the output signal from each line has the form ([9], p. 185)

$$u_{\text{out}}(t) = \frac{1}{\pi} \frac{\sin \frac{\Delta}{2}(t - T_k)}{t - T_k}\, e^{j[\omega_k t - l\gamma(\omega_k)]};$$

moreover,

$$T_k = l \left.\frac{d\gamma}{d\omega}\right|_{\omega=\omega_k} . \tag{7–3}$$

Therefore, strictly speaking, the total signal $s(t)$ must be written as follows:

$$s(t) = \frac{1}{\pi} \sum_{k=1}^{n} \beta_k \frac{\sin \frac{\Delta}{2}(t - T_k)}{t - T_k}\, e^{j[\omega_k t - l\gamma(\omega_k)]} . \tag{7–4}$$

For an ideal delay line, the time delay does not depend on frequency, and (7–4) reduces to (7–2).

The spectral component of the signal at the frequency ω_k is determined from the expression

$$S(\omega_k) = \int_{-\infty}^{\infty} s(t)\, e^{-j\omega_k t}\, dt .$$

Since the filter pass bands do not overlap, substituting (7–4) into the above results in the following single term for the sum spectral function:

$$S(\omega_k) = \frac{\beta_k}{\pi} e^{-l\gamma(\omega_k)} \int_{-\infty}^{\infty} \frac{\sin \frac{\Delta}{2}(t - T_k)}{t - T_k}\, dt = \beta_k\, e^{-jl\gamma(\omega_k)} .$$

This equation relates the spectrum of the signal being formed to the circuit parameters and, consequently, makes it possible to establish the basic relationships for the synthesis of the forming circuit. In fact, comparing this value of $S(\omega_k)$ with the required spectrum of the signal (7–1) and considering

(7–3), the needed values of β_k and T_k are found:

$$\beta_k = \sigma(\omega_k),$$

$$T_k = T_0' - \left.\frac{d\psi}{d\omega}\right|_{\omega=\omega_k}, \tag{7–5}$$

where T_0' is an arbitrary delay which is the same for all components; it characterizes the general delay of the signal introduced during its formation.

For any realizable circuit, the output signal cannot lead the input. The initial delay, T_0', must have such a value that T_k will be positive for all frequencies.

The number of transmitted data points must not be less than the number of degrees of freedom of the signal. Since two parameters, the amplitude and phase of the spectral component, are transmitted by each branch, the necessary number of branches is

$$n \geqslant FT = m,$$

where m is the compression ratio.

Consequently, the greater the compression ratio, the more complicated is the circuit required for forming the signal.

The circuit shown in Figure 11 can readily be modified so that instead of requiring many delay lines only one line with several branches is needed. In fact, it is possible to interchange the filtering and delay operation in all parallel branches, *i.e.*, to perform first the delay operation, then separate the spectral components, and finally perform the specified attenuation for each of the components. This is accomplished in the circuit shown in Figure 12.

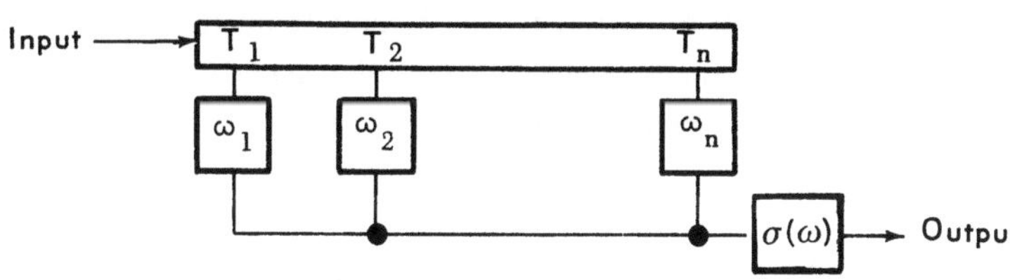

Fig. 12

The amplitudes of the spectral components are controlled after they are combined by connecting a filter with the characteristic $\sigma(\omega)$ to the summed output of all narrow-band filters.

Such a circuit can be used not only as a signal forming device but also for the optimal processing of signals in a receiver. In fact, the frequency characteristic of a matched filter,

$$H(\omega) = S^*(\omega) = \sigma(\omega)\,e^{-j\psi(\omega)},$$

differs from (7–1) only by the sign of $\psi(\omega)$. Therefore, to synthesize a matched filter, it is necessary only to change the delay time so that the sign of the phase shift is changed, *i.e.*, similar to (7–5)

$$\beta_k = \sigma(\omega_k),$$
$$T_k = T_0'' + \left.\frac{d\psi}{d\omega}\right|_{\omega=\omega_k}, \tag{7–6}$$

where T_0'' is once again an arbitrary delay which, of course, can be different from the value in the signal-forming circuit.

Forming and optimally processing a signal can be performed in one circuit. In addition to the obvious advantage of economy, such a procedure ensures automatic matching of the signal parameters to the optimal filter; any signal distortions appearing due to errors or instabilities in the forming circuits are automatically compensated for during the reception.

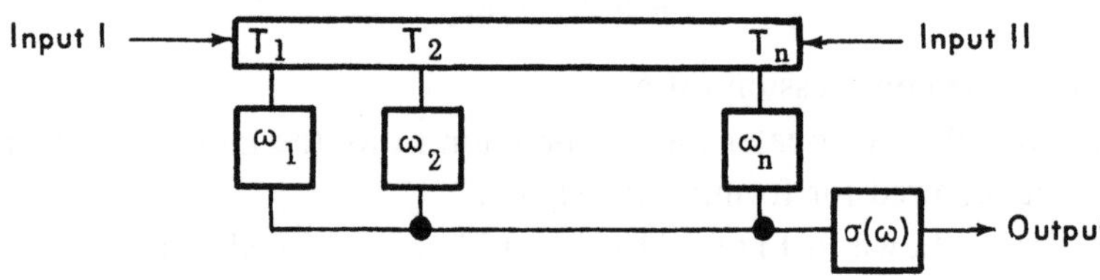

Fig. 13

Consider the circuit shown in Figure 13. This circuit differs from the previous one in that it has two inputs, and thus signals can be fed into the delay line from the left as well as from the right. To form a pulse, an input in the form of a δ-function is fed to the circuit from the left (Input I). A wide-band signal is formed at the output; the spectrum of the signal is defined by the frequency characteristic of the circuit:

$$S(\omega) = H_{\mathrm{lt}}(\omega) = \sigma(\omega)\, e^{j\psi_{\mathrm{lt}}(\omega)}.$$

From (7–5), the phase function $\psi_{\mathrm{lt}}(\omega)$ is determined from the condition

$$\frac{d\psi_{\mathrm{lt}}(\omega)}{d\omega} = T_0' - T(\omega)$$

or

$$\psi_{\mathrm{lt}}(\omega) = -\int_0^{\omega} T(\omega)\, d\omega + T_0'\omega + \psi_0'. \tag{7–7}$$

Here $T(\omega)$ is the delay as a continuous function of frequency, constructed from the discrete values $T_1, T_2, \ldots, T_n$.

It is now assumed that the input (δ-function) is fed into the circuit from the right (Input II). Frequencies which previously corresponded to long delays now correspond to short delays, and vice versa. Therefore, the frequency characteristic of the circuit for such a mode has the form

$$H_{\text{rt}}(\omega) = \sigma(\omega)\, e^{j\psi_{\text{rt}}(\omega)},$$

where the phase function $\psi_{\text{rt}}(\omega)$ differs from $\psi_{\text{lt}}(\omega)$ in that the sign of $T(\omega)$ is reversed, *i.e.*,

$$\frac{d\psi_{\text{rt}}(\omega)}{d\omega} = T_0'' + T(\omega),$$

$$\psi_{\text{rt}}(\omega) = \int_0^{\omega} T(\omega)\, d\omega + T_0''\omega + \psi_0''. \tag{7–8}$$

This circuit satisfies the requirements of (7–6) and, consequently, ensures the optimal processing of the signal that was formed prior to switching the inputs. Thus, for optimal processing of the signal formed, it is necessary to receive it via the other input of the delay line shown in Figure 13. The spectrum of the received pulse after processing is defined by the product

$$\begin{aligned} H_{\text{lt}}(\omega)\cdot H_{\text{rt}}(\omega) &= \sigma^2(\omega)\, e^{j[\psi_{\text{lt}}(\omega)+\psi_{\text{rt}}(\omega)]} \\ &= \sigma^2(\omega)\, e^{j[\omega(T'_0+T''_0)+\psi'_0+\psi''_0]}. \end{aligned}$$

It is readily seen that the resultant signal is the autocorrelation function having the phase of $\psi_0'+\psi_0''$ and delayed by the time $T_0'+T_0''$. Thus the circuit in Figure 13 accomplishes the optimal processing of the signal with the very same circuit that formed it, and the processing is independent of variation of the delays with respect to frequency. Therefore, instability of circuit parameters does not influence the quality of processing provided these parameters do not change over the short interval between the times of signal formation and reception.*

Signals formed at the low-power level are then amplified. Usually the phase structure does not change substantially with amplification (assuming a good amplifier design), but the amplitude modulation is almost completely eliminated. As a matter of fact, powerful SHF amplifiers (amplitrons, traveling-wave tubes, klystrons) operate under saturated conditions where the output level is defined not so much by the level of the input signal as by the driving power. To realize changes in the amplitude, the modulation must be accomplished at the final amplification stage and thus at the high-power level. Without going into the design details, it should be noted that a com-

* Considering this property, the circuit is sometimes called the "key-locked" circuit.

paratively flexible amplitude modulation cannot be accomplished under such conditions. The only practical possibility is switching the power amplifier on and off, and therefore the transmitted pulses are almost always close to rectangular.

Concerning the phase structure, two types of signals are basically used in practice: FM with the frequency changing monotonically, and phase modulation. In the latter case, a long-duration signal is composed of discrete sections, *i.e.*, rectangular pulses spaced close together. The carrier frequency is constant and the same for all sections, but the phase of the carrier changes with each section undergoing a step change at the boundaries. In its simplest form, the phase can have only one of two values over the pulse duration; these values differ by 180°. This means that the pulse has constant phase of either 0 or π, and that the order of the phase jumps defines the signal code. The spectrum width depends on the duration of the discrete section, and the product TF can be made very large.

2

The Uncertainty Principle

[§8] THE WOODWARD AMBIGUITY FUNCTION

It has thus far been assumed that the received signal is reflected from a stationary target so that the carrier frequency of the transmitted and reflected signals coincide. The general case will now be considered. Assume that the signal is reflected from a target whose range is r_0, and whose range rate is v_0. Thus the reflected signal is delayed by the time $t_0 = 2r_0/c$ and acquires a Doppler shift of $\Omega = (2v_0/c)\,\omega_0$:

$$s_1(t) = s(t - t_0)\, e^{j\Omega(t-t_0)}\,.$$

The signal spectrum is defined by the Fourier integral,

$$S_1(\omega) = \int_{-\infty}^{+\infty} s(t - t_0)\, e^{j\Omega(t-t_0)}\, e^{-j\omega t}\, dt = S(\omega - \Omega)\, e^{-j\omega t_0}\,.$$

If the signal is received by an optimal receiver with the frequency characteristic $S^*(\omega)$, the signal formed at the output is

$$y_s(t) = \frac{1}{2\pi} \int_{-\infty}^{\infty} S^*(\omega)\, S(\omega - \Omega)\, e^{j\omega(t-t_0)}\, d\omega\,. \tag{8–1}$$

This signal equation contains two unknowns: the time delay, t_0, and the Doppler frequency shift, Ω. The determination of these two parameters is a task for further processing the output signal, $y_s(t)$; the output signal is a function of the time delay t_0 and the Doppler frequency Ω, and is also a function of the radar which defines the accuracy and resolution of the range and range rate measurement. It is obvious that the measurement accuracy and resolution also depend on any subsequent processing and on the indication method. For example, if no provisions are made in the radar for determining the range rate, a portion of the information contained in the signal is not used. However, the potential of obtaining this target information is present in the signal $y_s(t)$.

Before proceeding further with the development, let us translate (8–1) by t_0 units so that $t=0$ is the time of the reflected signal's arrival. Then (8–1) becomes

$$y_s(t, \Omega) = \frac{1}{2\pi} \int_{-\infty}^{\infty} S^*(\omega)\, S(\omega - \Omega)\, e^{j\omega t}\, d\omega .$$

The function $|y_s(t, \omega)|$ is maximum when $t=0$ and $\Omega=0$; this follows from the results of §1. In fact, the receiver frequency characteristic $S^*(\omega)$ is matched to the received signal $s(t)$ when it is reflected from a stationary target ($\Omega=0$); in this case the maximum possible amplitude is obtained at the time $t=0$.

$$y_{\max} = y_s(0, 0) = \frac{1}{2\pi} \int_{-\infty}^{\infty} S^*(\omega)\, S(\omega)\, d\omega = E ,$$

where E is the total energy of the received signal.*

Therefore, it is possible to normalize the processed signal to the following form:

$$\frac{y_s(t, \Omega)}{y_{\max}} = \frac{1}{2\pi E} \int_{-\infty}^{\infty} S^*(\omega)\, S(\omega - \Omega)\, e^{j\omega t}\, d\omega .$$

Furthermore, by making a simple substitution of variables (*i.e.*, by shifting the frequency axis), the following symmetrical expression is obtained:

$$\frac{y_s(t, \Omega)}{y_{\max}} = \frac{e^{-j\frac{\Omega t}{2}}}{2\pi E} \int_{-\infty}^{\infty} S\left(\omega - \frac{\Omega}{2}\right) S^*\left(\omega + \frac{\Omega}{2}\right) e^{j\omega t}\, d\omega .$$

By dropping the unessential phase factor $e^{-j\Omega t/2}$, we obtain the so-called *ambiguity function* first derived by Woodward ([12], p. 120):

$$\chi(t, \Omega) = \frac{1}{2\pi E} \int_{-\infty}^{\infty} S\left(\omega - \frac{\Omega}{2}\right) S^*\left(\omega + \frac{\Omega}{2}\right) e^{j\omega t}\, d\omega . \tag{8–2}$$

It is clear from the previous discussion that the magnitude $|\chi(t, \Omega)|$ takes on the maximum value $|\chi|_{\max}=1$ at $t=\Omega=0$. Figure 14 is a three-dimensional plot showing $\chi(t, \Omega)$ above the t, Ω-plane.

The simplest way to explain the role of the ambiguity function is to compare it with the antenna pattern of the radar. The antenna pattern, $G(\varphi, \theta)$,

* See footnote on p. 3.

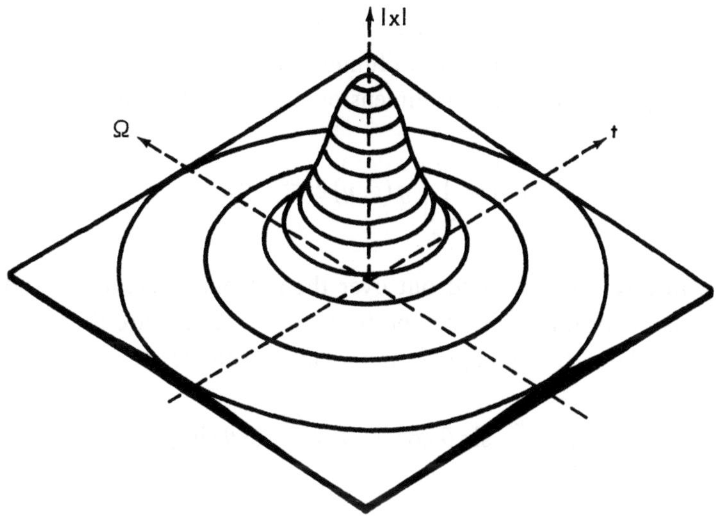

Fig. 14

fully defines the dependence of the received signal on the angular coordinates of a target. Here, φ and θ are the azimuth and elevation angles in the spherical coordinate system referenced to the antenna position. The antenna pattern, $G(\varphi, \theta)$, is a complex function which operates not only on the amplitude but also on the phase of the reflected signal. However, in most cases the dependence of the signal phase on the angular coordinates is not important, and only the amplitude characteristic, $|G(\varphi, \theta)|$, is considered. When probing for a target, the antenna pattern scans in an ordered fashion over the angular interval; locating the target coordinates is thus ensured.

The accuracy and resolution of the azimuth and elevation angle measurements depend substantially on the antenna pattern width in the various directions. Similar to the way the magnitude of the antenna pattern, $|G(\varphi, \theta)|$, defines the dependence of the received signal level on the angular direction, the magnitude of the ambiguity function, $|\chi(t, \Omega)|$, defines the dependence of the signal level on the target range (the delay time) and range rate (Doppler shift). *The narrower (sharper) the width of the ambiguity function in a given direction, the higher the accuracy and resolution of the range and range rate measurements if all other conditions are unchanged.*

Measurements of range and range rate depend on the shape of the ambiguity function to the same extent as angular position measurements depend on the shape of the antenna pattern. This explains why a great deal of attention has been given lately to the study of ambiguity functions and to the design of sophisticated signals that realize ambiguity functions with the desired form. There is also another analogy between ambiguity functions and antenna patterns. In designing an antenna, it is possible, more or less, to change arbitrarily the aperture illumination, $g(\xi, \eta)$, which is related to the antenna pattern by the two-dimensional Fourier transform (see Fradin [66], p. 86):

$$G(u, v) = \iint_{\sigma} g(\xi, \eta)\, e^{j(u\xi + v\eta)}\, d\xi\, d\eta . \tag{8–3}$$

Here the integration is carried out over the antenna aperture; ξ and η are the Cartesian coordinates in the aperture plane; u and v are the generalized coordinates related to the angular spherical coordinates:

$$u = \frac{2\pi}{\lambda} \sin\theta \cos\varphi, \quad v = \frac{2\pi}{\lambda} \sin\theta \sin\varphi ,$$

where λ is the wavelength.

Since the aperture illumination determines the two-dimensional pattern function, the problem is finding the proper aperture illumination to realize the desired antenna pattern. This is solved by evaluating the integral (8–3) for various $g(\xi, \eta)$.

The ambiguity function problem is solved in a similar manner. In designing a radar system, it is necessary to choose the signal $s(t)$ so that the desired ambiguity function is formed. It is not difficult to reduce this problem to an equation similar to (8–3). In fact, expressing $S^*(\omega + \Omega/2)$ in the integral form,

$$S^*\left(\omega + \frac{\Omega}{2}\right) = \int_{-\infty}^{\infty} s^*(t')\, e^{j\left(\omega + \frac{\Omega}{2}\right)t'}\, dt' ,$$

and substituting this value into (8–2), we obtain, after some simple transformations,

$$\chi(t, \Omega) = \frac{e^{j\frac{\Omega t}{2}}}{2\pi E} \int_{-\infty}^{\infty} \int_{-\infty}^{\infty} [S(\omega')\, s^*(t')\, e^{j\omega' t'}]\, e^{-j(t'\Omega - t\omega')}\, d\omega'\, dt' . \tag{8–4}$$

This relationship representing the ambiguity function in the form of the double Fourier integral is similar to (8–3). However, there is an essential difference between the above problems. The aperture illumination, $g(\xi, \eta)$, is an arbitrary function of the two variables ξ and η. The single limitation imposed on this function is that it may differ from zero only in the finite region bounded by the physical perimeter of the antenna. Therefore, any arbitrary function of the two variables u and v, limited to the possible values of the physical coordinates, may serve as an antenna pattern. Even such a comparatively minor constraint substantially limits the possible antenna patterns and complicates the synthesis problem.

The region of integration in (8–4) is essentially bounded also since it can be assumed that the signals have finite durations or spectrum widths.* However, there is an even more restrictive limitation. The product

$$S(\omega')\, s^*(t')\, e^{j\omega' t'}, \tag{8–5}$$

is similar to the function $g(\xi, \eta)$ in (8–4); this product consists of the signal $s(t')$ and its spectrum $S(\omega')$. When the form of the pulse is chosen, its spectrum becomes uniquely defined, and vice versa. Thus, the two-dimensional integrand used for determining the ambiguity function is not an arbitrary function of two variables, t' and ω', but a very special function in which one variable fully defines the other. This indicates that only a very narrow class of integrands can be used to realize ambiguity functions, and in the final analysis, leads to a number of important properties which define the restrictions when measuring simultaneously the range and range rate of a target. The principal property, the so-called *radar uncertainty principle*, is treated in the next section.

The following is another interpretation of the ambiguity function. For this purpose, it is necessary to change the order of integration in (8–4) and to integrate with respect to ω'. Then,

$$\frac{1}{2\pi}\int_{-\infty}^{\infty} S(\omega')\, e^{j\omega'(t+t')}\, d\omega' = s(t+t').$$

Consequently,

$$\chi(t, \Omega) = \frac{e^{j\frac{\Omega t}{2}}}{E}\int_{-\infty}^{\infty} s(t+t')\, s^*(t')\, e^{j\Omega t'}\, dt'$$

$$= \frac{1}{E}\int_{-\infty}^{\infty} s\left(t' + \frac{t}{2}\right) s^*\left(t' - \frac{t}{2}\right) e^{j\Omega t'}\, dt'. \tag{8–6}$$

Attention is drawn to the fact that (8–2) and (8–6) are very similar, and that the ambiguity function χ can be represented by two equivalent forms: in terms of the signal $s(t)$, or its spectrum $S(\omega)$. It should be noted that the intersection of the ambiguity function with the $\Omega = 0$ plane is essentially an autocorrelation function.

* This means that the integration is carried out over an infinite interval, the boundaries of which are parallel to one of the coordinate axes.

$$\chi(t, 0) = \frac{1}{E} \int_{-\infty}^{\infty} s\left(t' + \frac{t}{2}\right) s^*\left(t' - \frac{t}{2}\right) dt' = R(t).$$

Therefore, the ambiguity function is sometimes called the generalized autocorrelation function and also the signal function.

It is noted that by dropping the unessential phase factor it is also possible to represent the ambiguity function by the following equations:

$$\chi(t, \Omega) = \frac{1}{E} \int_{-\infty}^{\infty} s(t')\, s^*(t' - t)\, e^{j\Omega t'}\, dt', \tag{8–7}$$

$$\chi(t, \Omega) = \frac{1}{E} \int_{-\infty}^{\infty} s(t')\, s^*(t' + t)\, e^{j\Omega t'}\, dt', \tag{8–8}$$

and by analogous functions using the spectrum $S(\omega)$.

[§9] EXAMPLES

The ambiguity functions of some frequently encountered signals will now be calculated. First let us consider the Gaussian-shaped pulse:

$$s(t) = a_0\, e^{-\frac{t^2}{T^2}}\, e^{j\omega_0 t}. \tag{9–1}$$

Without loss of generality, the amplitude a_0 can be chosen so that the total energy of the signal is equal to unity (this normalization will also be used in the subsequent examples). Thus the signal is

$$s(t) = \left(\frac{2}{\pi T^2}\right)^{1/4} e^{-\frac{t^2}{T^2} + j\omega_0 t}.$$

Substituting this signal into (8–6), the following expression is obtained for the ambiguity function:

$$\chi(t, \Omega) = \sqrt{\frac{2}{\pi}}\frac{1}{T}\, e^{-\frac{t^2}{2T^2} + j\omega_0 t} \int_{-\infty}^{\infty} e^{-2\frac{t'^2}{T^2} + j\Omega t'}\, dt'$$

$$= e^{-\frac{1}{2}\left[\frac{t^2}{T^2} + \frac{\Omega^2}{(2/T)^2}\right]}\, e^{j\omega_0 t}. \tag{9–2}$$

It can be seen from (9–2) that the absolute value of the ambiguity function $|\chi(t, \Omega)|$ has a plot in the form of a hill as is shown in Figure 14. The peak of the hill is at the origin $t=\Omega=0$, with the maximum value of $|\chi|$ being unity.

Ambiguity functions (more precisely their magnitudes) are frequently drawn in the form of contour lines, similar to the technique used in topography, to simplify the plotting. For this example (see Fig. 15) these contours

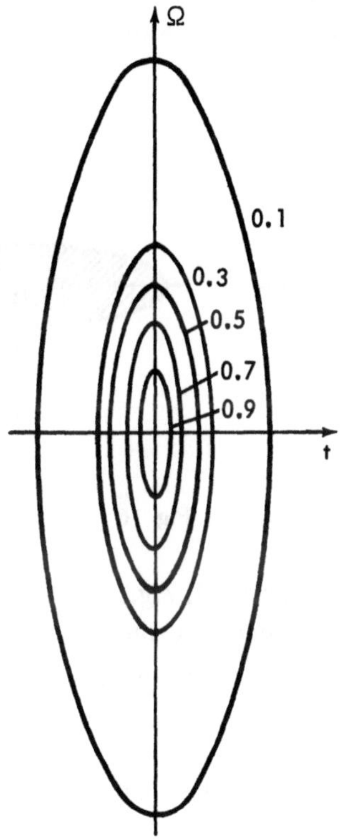

Fig. 15

are ellipses given by equation

$$\frac{t^2}{T^2} + \frac{\Omega^2}{(2/T)^2} = c^2 . \tag{9–3}$$

The constant c determines the height that the ambiguity function contour represents. It follows from (9–2) that

$$\tfrac{1}{2}c^2 = -\ln |\chi(t, \Omega)| .$$

It can readily be seen that the area of each ellipse does not depend on the pulse duration T. In fact, the area of an ellipse is proportional to the product of its semi-axes, which in this case is

$$\pi(cT)\left(c\frac{2}{T}\right) = 2\pi c^2 = -4\pi \ln|\chi|.$$

Figure 16 shows the cross sections of the ambiguity functions for two types of signals: (a) is for short pulses and (b) is for long pulses; while the scale is arbitrary, it is the same for both drawings. These cross sections are

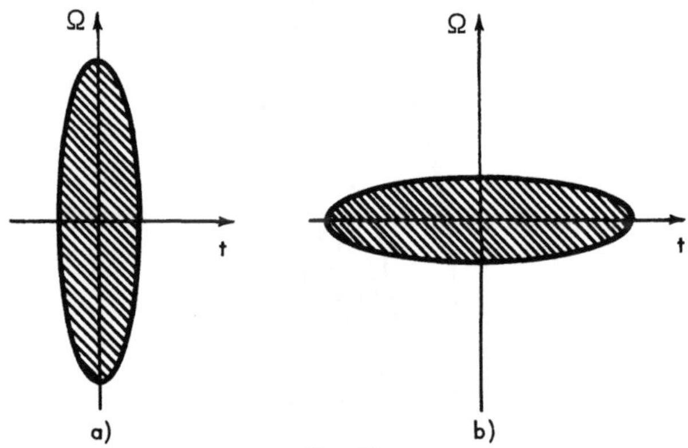

Fig. 16

drawn for the same contour level, *e.g.*, the 0.5 level (this is the common practice). In example (a), where the pulse duration is short, the ellipse is compressed along the time axis and elongated along the frequency axis. In example (b), the ellipse is elongated along the time axis and compressed along the frequency axis. Moreover, the areas of the ellipses (and the product of their semi-axes) are the same. It was noted in the previous section that the width of the ambiguity function along the axes is usually related to accuracy and resolution of the range and range rate. It can tentatively be assumed that the differences in range and range rates between targets cannot be discriminated if their points (in the t, Ω-plane) lie inside the shaded ellipse. Similarly, it can be assumed that two targets lying inside the ellipse cannot be resolved. It follows that by changing the duration of a Gaussian-shaped pulse it is possible to gain in the accuracy and resolution of the range measurement, but at the same time the accuracy and resolution of the range rate measurements are reduced, or vice versa. Here for the first time we encounter the *uncertainty principle* in radar, according to which there exists a certain "unique" uncertainty (ambiguity) in determining the target range and range rate simultaneously. By changing the signal parameters, it is only possible to redistribute this uncertainty (*i.e.*, a gain in range accuracy at the expense of range rate, or vice versa), but not to decrease it. A rigorous proof and interpretation of this principle is given in the next section.

For the second example, a rectangular pulse having unity energy and with time duration T is considered.

$$s(t) = \begin{cases} \dfrac{1}{\sqrt{T}} e^{j\omega_0 t} & \text{for} \quad |t| \leqslant \dfrac{T}{2}, \\ 0 & \text{for} \quad |t| > \dfrac{T}{2}. \end{cases}$$

The spectrum of such a signal has the form

$$S(\omega) = \frac{2}{\sqrt{T}} \frac{\sin(\omega - \omega_0)\dfrac{T}{2}}{\omega - \omega_0}.$$

Using (8–2), the ambiguity function is calculated to be

$$\chi(t, \Omega) = \frac{2}{\pi T} \int_{-\infty}^{\infty} \frac{\sin\left(\omega - \omega_0 - \dfrac{\Omega}{2}\right)\dfrac{T}{2}}{\omega - \omega_0 - \dfrac{\Omega}{2}} \times \frac{\sin\left(\omega - \omega_0 + \dfrac{\Omega}{2}\right)\dfrac{T}{2}}{\omega - \omega_0 + \dfrac{\Omega}{2}} e^{j\omega t} \, d\omega .$$

This integral has been evaluated by Gradshtein and Ryzhik ([17], pp. 430, 421). Using this value, one finds, after some simple transformations,

$$\chi(t, \Omega) = \begin{cases} \dfrac{\sin\left(1 - \dfrac{|t|}{T}\right)\dfrac{\Omega T}{2}}{\Omega T/2} e^{j\omega_0 t} & \text{for} \quad |t| \leqslant T, \\ 0 & \text{for} \quad |t| > T. \end{cases} \tag{9–4}$$

The absolute value of the ambiguity function, $|\chi(t, \Omega)|$, has its maximum at $t = \Omega = 0$; moreover, $\chi(0, 0) = 1$. The cross sections of the ambiguity function through the $\Omega = 0$ and $t = 0$ planes, and also the horizontal cross sections for two pulse durations are illustrated in Figure 17 (a) and (b), respectively. It is not difficult to show that for sufficiently high levels of $|\chi|$, the contours can be approximately expressed by the following equation:

$$\frac{|t|}{T} + \frac{\Omega^2}{6(2/T)^2} = c^2 .$$

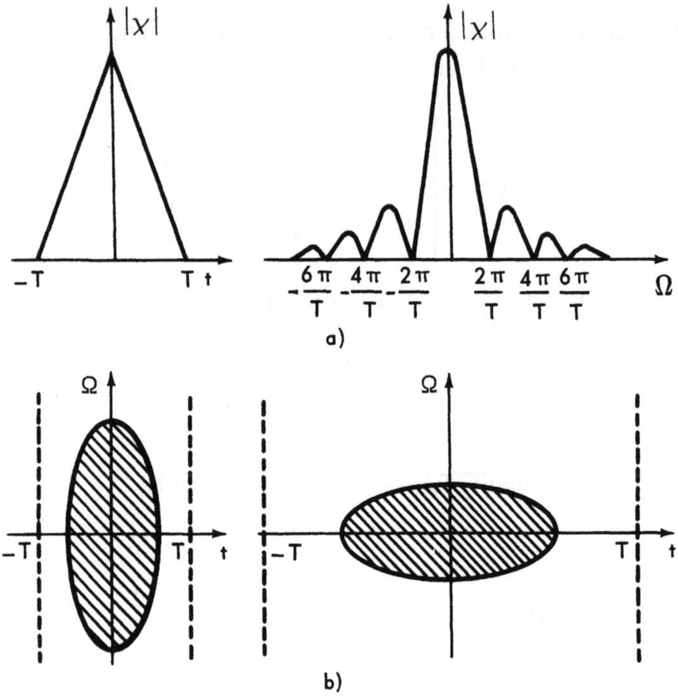

Fig. 17

Obviously, the same qualitative law mentioned previously also holds for rectangular pulses: by shortening the pulse duration, the accuracy and resolution of the range improves but becomes worse for the range rate.

Let us now consider a coherent pulse train consisting of $N+1$ pulses having identical amplitudes (see Fig. 18). The mathematical equation for

Fig. 18

this signal is

$$s(t) = \sum_{k=0}^{N} s_0(t - kT_1),$$

where $s_0(t)$ is the expression for a single pulse.

The pulses are of short duration and do not overlap; the total signal energy equals the sum of the energies of the individual pulses:

$$E = (N + 1) E_0 = 1 .$$

Normalizing the amplitude, the equation of the signal becomes

$$s(t) = \frac{1}{\sqrt{(N+1)\,E_0}} \sum_{k=-\infty}^{\infty} \alpha_k s_0(t - kT_1). \tag{9–5}$$

For convenience, the amplitude factor α_k defined below is introduced (k can only assume integer values):

$$\alpha_k = \begin{cases} 1 & \text{for} \quad 0 \leqslant k \leqslant N, \\ 0 & \text{for} \quad k < 0, k > N. \end{cases}$$

Thus the summation can be carried out over infinite limits.

The spectrum of the signal (9–5) is determined by the Fourier integral

$$\begin{aligned} S(\omega) &= \frac{1}{\sqrt{(N+1)\,E_0}} \sum_k \alpha_k \int_{-\infty}^{\infty} s_0(t - kT_1)\, \mathrm{e}^{-j\omega t}\, dt \\ &= \frac{S_0(\omega)}{\sqrt{(N+1)\,E_0}} \sum_k \alpha_k\, \mathrm{e}^{-jk\omega T_1}, \end{aligned}$$

where $S_0(\omega)$ is the spectrum of the individual pulses $s_0(t)$.

Substituting this spectrum into (8–2), we obtain

$$\begin{aligned} \chi(t, \Omega) &= \frac{1}{N+1} \sum_k \sum_l \alpha_k \alpha_l\, \mathrm{e}^{j(k+l)\frac{\Omega T_1}{2}} \\ &\quad \times \frac{1}{2\pi E_0} \int_{-\infty}^{\infty} S_0\left(\omega - \frac{\Omega}{2}\right) S_0^*\left(\omega + \frac{\Omega}{2}\right) \mathrm{e}^{j\omega[t-(k-l)T_1]}\, d\omega \\ &= \frac{1}{N+1} \sum_k \sum_l \alpha_k \alpha_l\, \mathrm{e}^{j(k+l)\frac{\Omega T_1}{2}} \chi_0[t - (k-l)\, T_1, \Omega], \end{aligned}$$

where χ_0 is the ambiguity function of the individual pulse.

Let us introduce into the double sum the new summation index

$$n = k - l, \; -N \leqslant n \leqslant N.$$

Then the ambiguity function becomes

$$\chi(t, \Omega) = \sum_{n=-N}^{N} \chi_0(t - nT_1, \Omega)\, Q_n(\Omega), \tag{9–6}$$

where

$$Q_n(\Omega) = \frac{1}{N+1}\, \mathrm{e}^{jn\frac{\Omega T_1}{2}} \sum_l \alpha_l \alpha_{l+n}\, \mathrm{e}^{jl\Omega T_1}.$$

The second summation is extended over all l for which the product $\alpha_l \cdot \alpha_{l+n}$ differs from zero (equals unity). It is readily seen that these values are defined by the conditions

$$0 \leqslant l \leqslant (N-n) \quad \text{for} \quad n>0,$$
$$|n| \leqslant l \leqslant N \quad \text{for} \quad n<0.$$

Furthermore, the summation is a geometric progression that can be totalized in closed form. As a result,

$$Q_n(\Omega) = \frac{\sin(N-|n|+1)\dfrac{\Omega T_1}{2}}{(N+1)\sin\dfrac{\Omega T_1}{2}} e^{jn\frac{T_1}{2}}, \qquad |n| \leqslant N.$$

If the number of pulses in the pulse train is large, this function has sharp peaks at the values of Ω for which $\Omega T_1 = 2m\pi$. For these values

$$Q_{nm} = \frac{N-|n|+1}{N+1}(-1)^{nm}.$$

Figure 19 is a contour plot of this ambiguity function. The dotted curves spaced sequentially along the time axis correspond to the ambiguity functions for the individual pulses, $\chi_0(t-nT_1, \Omega)$. These ambiguity functions are additionally "modulated" by the factor $Q_n(\Omega)$. The resulting function, χ, has a periodic structure in both directions. It is not difficult to interpret the character of such an ambiguity function. The time periodicity is indicative of the ambiguity in the range measurement which appears in cases where the target's range (in time units) is greater than the pulse repetition period. This

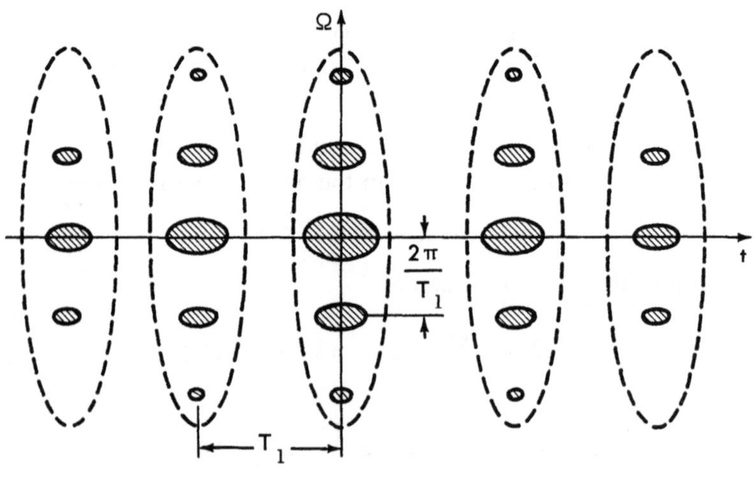

Fig. 19

is a well-known phenomenon, and in designing radar systems the period T_1 is usually chosen greater than the radar range to remove the ambiguity of the range measurement. The periodicity (ambiguity) with respect to frequency is also known for systems that select moving targets. This ambiguity is related to the so-called "blind" range rates of the targets. If the repetition period, T_1, and the carrier frequency, ω_0, are chosen such that the ambiguity does not manifest itself within the expected limits of range and range rate, the accuracy and resolution are defined only by the central peak of the function χ. The width of the central pattern is substantially smaller than that for a single pulse signal, and thus the coherent pulse train has definite practical advantages.* However, the use of a pulse train instead of a single pulse does not remove the general ambiguity of the measurement but redistributes it; a set of ambiguous lobes is obtained, each of which has a higher accuracy.

For the last example, let us consider a frequency-modulated pulse having a Gaussian envelope discussed in §3. Normalizing the energy as was done previously, the signal has the form

$$s(t) = \left(\frac{2}{\pi T^2}\right)^{1/4} \mathrm{e}^{-\frac{t^2}{T^2}} \mathrm{e}^{j(\omega_0 t + \gamma t^2)} .$$

Substituting this signal into (8–6) and performing some simple but cumbersome operations, the following expression for the ambiguity function is obtained:

$$\chi(t, \Omega) = \mathrm{e}^{-\frac{1}{2}\left[m^2 \frac{t^2}{T^2} + (\gamma T^2)\,\Omega t + \frac{\Omega^2}{(2/T)^2}\right]} \mathrm{e}^{j\omega_0 t} , \tag{9–7}$$

where $m = \sqrt{1 + \gamma^2 T^4}$ is the compression ratio.

Here the height contours are defined by

$$m^2 \frac{t^2}{T^2} + (\gamma T^2)\,\Omega t + \frac{\Omega^2}{(2/T)^2} = c^2 .$$

As in the case of a Gaussian-shaped pulse without frequency modulation, the constant c depends only on the contour level:

$$\tfrac{1}{2} c^2 = -\ln |\chi(t, \Omega)| .$$

It should also be noted that when $\gamma = 0$, (9–7) becomes (9–2), *i.e.*, the ambiguity function of a pulse with constant frequency is obtained.

In order to explain the nature of the ambiguity function of a frequency-modulated signal, we shall introduce the dimensionless coordinates which

* To realize these advantages, it is of course necessary to use a receiver matched to the pulse train, *i.e.*, a comb filter is needed.

are proportional to time and frequency:

$$\xi = \frac{t}{T}, \quad \eta = \frac{\Omega T}{2}.$$

Then the equation for the height contours takes the form

$$m^2 \xi^2 + 2(\gamma T^2)\, \xi\eta + \eta^2 = c^2. \tag{9–8}$$

With these coordinates, the height contour for a pulse without frequency modulation is a circle [*cf.* (9–3)].

$$\xi^2 + \eta^2 = c^2.$$

It is readily seen that the circle transforms into an ellipse when the pulse has frequency modulation. It can be shown, using the well-known rules for writing the equation in terms of the axes intercepts, that the conical form of (9–8) is

$$\frac{\xi_1^2}{a^2} + \frac{\eta_1^2}{b^2} = c^2.$$

The semi-axes of the ellipse are equal to a and b; moreover,

$$\frac{1}{a^2} = \tfrac{1}{2}(m^2+1)\left(1 + \sqrt{1 - \frac{4}{(m^2+1)^2}}\right) \approx m^2 + 1 \approx m^2,$$

$$\frac{1}{b^2} = \tfrac{1}{2}(m^2+1)\left(1 - \sqrt{1 - \frac{4}{(m^2+1)^2}}\right) \approx \frac{1}{m^2+1} \approx \frac{1}{m^2}.$$

Thus one semi-axis of the ellipse is compressed by the factor m, and the other is elongated by the factor m. However, the area of the ellipse does not change. It can also be shown that the major axis of the ellipse is inclined to the ξ-axis (time axis) at the angle φ; moreover,

$$\tan\varphi = \left[\frac{\sqrt{4+\gamma^2T^4} + \gamma T^2}{\sqrt{4+\gamma^2T^4} - \gamma T^2}\right]^{1/2} \approx \gamma T^2\left(1 + \frac{1}{2\gamma^2T^4}\right) \approx \gamma T^2.$$

The equation of the ellipse's major axis in the ξ and η coordinates has the form

$$\eta \approx \gamma T^2 \xi.$$

Returning to the initial t, Ω coordinate system, this equation takes the following form:

$$\Omega \approx 2\gamma t,$$

i.e., the direction of the ellipse's major axis is defined by the rate of change of the signal's instantaneous frequency. It is noted that in these equations it was assumed that the compression ratio is large, *i.e.*,

$$m^2 \approx \gamma^2 T^4 \gg 1 .$$

The change in the ambiguity function caused by the introduction of the frequency modulation is shown in Figure 20. The shaded ellipse corresponding to frequency modulation cuts off at the small segment T/m on the time

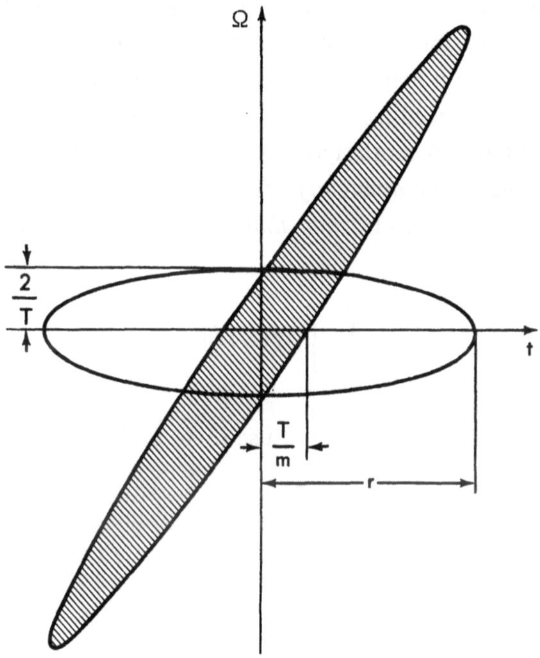

Fig. 20

axis; this indicates that in compressing the signal by a factor m, the accuracy and resolution of the range measurement are improved. The accuracy and resolution of the range rate measurement remains unchanged (both ellipses intersect the Ω-axis at the same point). However, the overall uncertainty of the frequency and time measurements does not improve since the area of the ellipse remains the same.

[§10] SOME PROPERTIES OF AMBIGUITY FUNCTIONS—THE RADAR UNCERTAINTY PRINCIPLE

A general rule can be drawn from the examples discussed in the previous section. For the various signal forms (*e.g.*, Gaussian-shaped pulses of various durations, with and without frequency modulation), the overall uncertainty of the range and range rate measurements is preserved, since the areas of the function contours for a given level (ellipses) are the same. Qualitatively, similar results are obtained also for such signals as the coherent pulse train; the pulse train is, seemingly, so different from the single pulse that a substantial departure from the above rule could be expected.

This suggests that there is an inherent invariance in the product of the range and range rate measurement accuracy and resolution; by changing the signal form, it is possible to change the accuracy of the range and range rate measurement in such a manner that a gain for one parameter leads to a loss for the other. The preceding statement constitutes the essence of the radar uncertainty principle. A more rigorous statement of the uncertainty principle is that the *volume of the ambiguity "body" (i.e., the volume bounded by the function* $|\chi(t, \Omega)|^2$*) is an invariant that does not depend on the signal form.* It will be shown that

$$V = \frac{1}{2\pi} \int_{-\infty}^{\infty} \int_{-\infty}^{\infty} |\chi(t, \Omega)|^2 \, dt \, d\Omega = 1 . \tag{10–1}$$

To prove (10–1) it is necessary to use (8–6) and thus,

$$\begin{aligned} |\chi(t, \Omega)|^2 &= \chi(t, \Omega)\, \chi^*(t, \Omega) \\ &= \frac{1}{E^2} \int_{-\infty}^{\infty} s\left(t' + \frac{t}{2}\right) s^*\left(t' - \frac{t}{2}\right) e^{j\Omega t'} \, dt' \\ &\times \int_{-\infty}^{\infty} s^*\left(t'' + \frac{t}{2}\right) s\left(t'' - \frac{t}{2}\right) e^{-j\Omega t''} \, dt'' . \end{aligned}$$

Substituting the above into (10–1) and changing the order of integration, we obtain

$$\begin{aligned} V = \frac{1}{2\pi E^2} \iiint_{-\infty}^{\infty} & s\left(t' + \frac{t}{2}\right) s^*\left(t' - \frac{t}{2}\right) s^*\left(t'' + \frac{t}{2}\right) \\ & \times s\left(t'' - \frac{t}{2}\right) dt' \, dt'' \, dt \int_{-\infty}^{\infty} e^{j\Omega(t' - t'')} \, d\Omega . \end{aligned}$$

The last integral is the δ-function, *i.e.*,

$$\frac{1}{2\pi} \int_{-\infty}^{\infty} e^{j\Omega(t' - t'')} \, d\Omega = \delta(t' - t'') .$$

Therefore,

$$V = \frac{1}{E^2} \iiint_{-\infty}^{\infty} s\left(t' + \frac{t}{2}\right) s^*\left(t' - \frac{t}{2}\right) s^*\left(t'' + \frac{t}{2}\right)$$

$$\times s\left(t'' - \frac{t}{2}\right) \delta(t' - t'')\, dt''\, dt'\, dt\,.$$

Furthermore, by integrating with respect to t'' and using the basic property of the δ-function, we find

$$V = \frac{1}{E^2} \iint_{-\infty}^{\infty} \left|s\left(t' + \frac{t}{2}\right)\right|^2 \cdot \left|s\left(t' - \frac{t}{2}\right)\right|^2 dt'\, dt$$

$$= \frac{1}{E^2} \int_{-\infty}^{+\infty} |s(x)|^2\, dx \cdot \int_{-\infty}^{\infty} |s(y)|^2\, dy\,.$$

Each of the integrals in the above is equal to the signal energy, E; consequently, $V=1$, and (10–1) is proved.

This relationship is valid for a signal having any arbitrary form, duration, spectrum width, etc. Using the figurative comparison of Siebert [28], it can be said that "the ambiguity function is the surface of a pile of sand. As we adjust the waveform, we seem to be able to move the sand around but unable to get rid of any of it."

It is important to note that due to the normalization, the height of the ambiguity function is equal to unity at its maximum, $\chi(0, 0)$. Therefore, by compressing the ambiguity function along the time axis, it will unavoidably spread along the frequency axis; while forming a narrow peak of the ambiguity function at the origin, it is necessary to distribute somehow the remaining volume of the "heap" over the t, Ω-plane (perhaps in the form of other peaks placed on the periphery). This is the case of the coherent pulse train.

Let us now introduce the concept of the *effective cross section of the ambiguity function*. For this purpose, it is necessary to construct the three-dimensional plot of the function $|\chi(t, \Omega)|^2$ above the t, Ω-plane and consider the cross sections of the function at various heights of $h \leqslant 1$. Denoting the area of each cross section by $F(h)$, we can write (10–1) as

$$V = \frac{1}{2\pi} \int_0^1 F(h)\, dh = 1\,.$$

From the theorem of the mean, there exists a height h_{eff} for which

$$\int_0^1 F(h)\,dh = F(h_{\text{eff}}), \quad 0 < h_{\text{eff}} < 1 .$$

The effective cross section of the ambiguity function is the cross section of this function at the height h_{eff}. According to this definition, a cylinder with unity height and whose cross section equals the effective cross section (such a volume may consist of several separate cylinders) has a volume equal to that of the function $|\chi|^2$. Such a definition is similar in many respects to that used in §4 for the effective duration of a pulse (*i.e.*, the duration of a rectangular pulse having the equivalent amplitude and energy). In the previous section, an arbitrary cross section of the χ-function was used. The equality of the cross sections was established only for certain cases (Gaussian-shaped pulses with or without frequency modulation). The characteristic feature of these cases is that the cross sections are equal at any (for the different signals) level.

Using the concept of effective cross section, the radar uncertainty principle can be stated as follows: *the area of the effective cross section of the function $|\chi|^2$ does not depend on the form of the signal and equals 2π.*

The analogy between the antenna pattern and the ambiguity function was discussed in §8. It was noted there that, by virtue of the special nature of the integrand of (8–4), the class of admissible ambiguity functions is more restrictive than that for antenna patterns. Now the statement can be defined more accurately. By increasing the antenna dimensions, it is possible arbitrarily to narrow down the beam and thus reduce the cross-sectional area of the pattern to any given small value. However, for the ambiguity function, the situation is different: no matter how much the "dimensions" of the signal are changed (*i.e.*, the pulse duration and spectrum width), the effective cross section of the function $|\chi|^2$ remains the same; it is not possible to obtain an ambiguity function χ with an arbitrarily narrow beam.

Eq. (10–1) which establishes the invariance of the ambiguity function volume may be interpreted as a special case of a more general relationship that will be needed for the following development. To prove this, some definitions are needed.

Let us take two different signals, $s_1(t)$ and $s_2(t)$, and form the *cross-ambiguity function,*

$$\chi_{1,2}(t, \Omega) = \frac{1}{\sqrt{E_1 E_2}} \int_{-\infty}^{\infty} s_1\left(t' + \frac{t}{2}\right) s_2^*\left(t' - \frac{t}{2}\right) e^{j\Omega t'}\, dt' . \qquad (10\text{–}2)$$

This quantity is similar to the cross-correlation function and is of great importance in communication theory. If a communication channel is intended for transmitting a number of signals $s_i(t)\,(i=1, 2, 3, \ldots)$, then in the optimal case the discrimination between signals at the receiving end reduces to the comparison of the type of functions as defined by (10–2). Similarly, let us form the cross-ambiguity function $\chi_{3,4}(t, \Omega)$ for two other signals $s_3(t)$ and $s_4(t)$, and take the two-dimensional Fourier transform of the product $\chi_{1,2}\cdot\chi_{3,4}^*$:

$$\varphi(t_1, \Omega_1) = \frac{1}{2\pi} \int_{-\infty}^{\infty} \int_{-\infty}^{\infty} \chi_{1,2}(t, \Omega)\, \chi_{3,4}^*(t, \Omega)\, e^{j(\Omega t_1 - t\Omega_1)}\, dt\, d\Omega\,. \tag{10–3}$$

This integral can be evaluated in a similar fashion as was done in deriving (10–1). Making use of (10–2), the product $\chi_{1,2}\cdot\chi_{3,4}^*$ in (10–3) is in the form of a double integral. Substituting this double integral into (10–3), the resulting function $\varphi(t_1, \Omega_1)$ is in the form of a quadruple integral. Changing the order of integration and making use of the δ-function properties, we find

$$\varphi(t_1, \Omega_1) = \frac{1}{\sqrt{E_1 E_2 E_3 E_4}} \iint_{-\infty}^{\infty} s_1\left(t' + \frac{t}{2}\right) s_2^*\left(t' - \frac{t}{2}\right)$$
$$\times\, s_3^*\left(t' + t_1 + \frac{t}{2}\right) s_4\left(t' + t_1 - \frac{t}{2}\right) e^{j\Omega_1 t}\, dt'\, dt\,.$$

The use of further coordinate transformations finally leads to [56]

$$\varphi(t_1, \Omega_1) = \frac{1}{2\pi} \int_{-\infty}^{\infty} \int_{-\infty}^{\infty} \chi_{1,2}(t, \Omega)\, \chi_{3,4}^*(t, \Omega)\, e^{j(\Omega t_1 - t\Omega_1)}\, dt\, d\Omega$$
$$= \chi_{4,2}(t_1, \Omega_1)\, \chi_{3,1}^*(t_1, \Omega_1)\,. \tag{10–4}$$

Assuming that all the signals are identical, *i.e.*, $s_1 = s_2 = s_3 = s_4 = s(t)$, the following important relationship is obtained:

$$\frac{1}{2\pi} \int_{-\infty}^{\infty} \int_{-\infty}^{\infty} |\chi(t, \Omega)|^2\, e^{j(\Omega t_1 - t\Omega_1)}\, dt\, d\Omega = |\chi(t_1, \Omega_1)|^2\,. \tag{10–5}$$

Thus the two-dimensional Fourier transform of the function $|\chi(t, \Omega)|^2$ is equal to the function itself. It can be stated, using the terminology of Fourier integral theory, that the function $|\chi(t, \Omega)|^2$ *is self-reciprocal in the two-dimensional Fourier transform.*

Eq. (10–5) reduces to the radar uncertainty principle when the χ-function is normalized and t_1, Ω_1 are set to zero. Also, note that the ambiguity function has the following symmetry:

$$\chi(t, \Omega) = \chi^*(-t, -\Omega). \tag{10–6}$$

This relationship can be readily proved by using either (10–2) or (8–6). Accordingly, the symmetry of $|\chi(t, \Omega)|$ is

$$|\chi(t, \Omega)| = |\chi(-t, -\Omega)|.$$

As was noted previously, the "sharpness" of the ambiguity function in the t or Ω direction is of great importance to the measurement accuracy of the target parameter. It can be readily shown that this sharpness is characterized by the derivatives $\partial^2\chi/\partial t^2$ and $\partial^2\chi/\partial\Omega^2$ at the central point of $t=\Omega=0$. Moreover, the larger (in absolute value) these derivatives, the narrower the ambiguity function in the corresponding direction. From (10–2) and (8–6) it follows that

$$\left.\frac{\partial^2\chi}{\partial t^2}\right|_{t=\Omega=0} = -\frac{1}{2\pi E}\int_{-\infty}^{\infty} \omega^2 |S(\omega)|^2 \, d\omega = -4\pi^2 F^2, \tag{10–7}$$

$$\left.\frac{\partial^2\chi}{\partial \Omega^2}\right|_{t=\Omega=0} = -\frac{1}{E}\int_{-\infty}^{\infty} t^2 |s(t)|^2 \, dt = -T^2, \tag{10–8}$$

where T and F are the "root mean square" duration and spectrum width of the signal.

These quantities are discussed in greater detail in §12; however, it is important to emphasize here that the width of the ambiguity function in the time direction depends on the spectrum of the signal, and the width in the frequency direction depends on the signal pulse duration.

The derivative $\partial^2\chi/\partial t\,\partial\Omega$ evaluated at the central point is determined by the following method. First of all,

$$\left.\frac{\partial^2}{\partial t\,\partial\Omega} e^{j\left(\omega' t + t'\Omega + \frac{\Omega t}{2}\right)}\right|_{t=\Omega=0} = \frac{j}{2} + \omega' t'.$$

Differentiating (8–4) and using the above, we obtain

$$\left.\frac{\partial^2\chi}{\partial t\,\partial\Omega}\right|_{t=\Omega=0} = \frac{j}{2}\chi(0, 0) - \frac{1}{2\pi E}\iint_{-\infty}^{\infty} t\omega S(\omega)\, s^*(t)\, e^{j\omega t}\, dt\, d\omega$$

$$= \frac{j}{2} + \frac{j}{E}\int_{-\infty}^{\infty} t s'(t)\, s^*(t)\, dt.$$

The signal expressed in terms of its amplitude and phase functions is

$$s(t) = a(t)\, e^{j\varphi(t)}$$

Substituting this signal function into the integral above and integrating by parts,

$$\int_{-\infty}^{\infty} t s'(t)\, s^*(t)\, dt = \int_{-\infty}^{\infty} t a'(t)\, a(t)\, dt + j \int_{-\infty}^{\infty} t\varphi'(t)\, a^2(t)\, dt = -\frac{E}{2} + j \int_{-\infty}^{\infty} t\varphi'(t)\, |s(t)|^2\, dt .$$

Finally,

$$\left.\frac{\partial^2 \chi}{\partial t\, \partial \Omega}\right|_{t=\Omega=0} = -\frac{1}{E} \int_{-\infty}^{\infty} t\varphi'(t)\, |s(t)|^2\, dt . \qquad (10\text{–}9)$$

Evidently, this derivative with respect to both variables is a measure of the signal's frequency modulation. Roughly speaking, it is the phase advance acquired over the pulse duration. If the signal frequency is constant, then this derivative equals zero.

[§11] ANALOGY WITH QUANTUM PHYSICS

One of the basic tenets of quantum physics is the Heisenberg uncertainty principle. According to this principle, not all physical quantities observed in any realizable experiment (even in principle only) can be determined with an arbitrarily high accuracy. There are mutually exclusive (conjugate, non-commutating) quantities the measurement errors of which are interrelated. Even under ideal experimental conditions, an increase in the measurement accuracy of one quantity can be achieved (after a certain limit) only at the expense of decreasing the measurement accuracy of another quantity.

The coordinate of a particle, x, and its momentum in the x-direction,

$$p_x = m\frac{dx}{dt} = mv_x ,$$

is one example of two such mutually exclusive quantities.

The measurement accuracies of these two quantities have the following uncertainty relationship:

$$\delta x \delta p_x \geqslant \frac{h}{4\pi} = \frac{\hbar}{2}, \qquad (11\text{–}1)$$

where $h = 6.625 \times 10^{-34}$ *joule-sec* (Planck's constant).

Since the momentum p_x is proportional to the velocity v_x, it follows from the uncertainty principle that it is not possible to determine experimentally the position and velocity of a particle simultaneously. This in turn means that the motion of a single particle cannot be characterized by its trajectory as it is understood in the usual sense.

From the previous discussion, it can be seen that there is a similar situation with radar measurements. Radar observations can be interpreted as an experiment for determining the target coordinates (range) and range rate at a given instant of time. The properties of the χ-function are such that the measurement accuracy of the range and range rate are interrelated; the improvement of the range accuracy results in the worsening of the range rate accuracy, and vice versa. It was not by chance that the terms "measurement uncertainty," "uncertainty principle," etc., were transferred to radar terminology from quantum physics. However, it is necessary to distinguish between the physical nature of uncertainty in the micro-world and radar. With this purpose in mind, we shall dwell briefly upon some principles of quantum mechanics.

As with any new physical theory, quantum mechanics was developed to explain a number of phenomena contrary to classical physics that were observed experimentally. These experimental facts can be classified into two groups.

(1) In experiments involving the photoelectric effect, emission spectra, gas absorption, X-ray scattering, etc., it was observed that electromagnetic energy is transferred in elementary units, or quanta. Every energy quantum (photon) is emitted or absorbed as a whole; it is possible to substitute one quantum by another, but it is not possible to reduce a quantum into smaller units. The magnitude of the quantum (energy of one photon) is proportional to the emission frequency, ν, and is defined by the relationship

$$E = h\nu = \hbar\omega\,. \tag{11–2}$$

Also, all electromagnetic phenomena have a wave nature manifested by interference and diffraction. At high intensities, these effects are in full agreement with the classical theory which treats emission as being a continuous process and reduces the electromagnetic phenomena to simple oscillations, *i.e.*, waves.

(2) In bombarding crystals by electrons, a diffraction of these electrons was observed in the crystalline lattice; this diffraction is similar to that of electromagnetic and other waves. The observation shows that the motion of an electron is accompanied by some wave process. Subsequently, diffraction phenomena were also observed in experiments involving heavier particles (including neutrons which do not have an electrical charge). This

diffraction process can be explained quantitatively if it is assumed that the particle momentum, p, is related to the oscillation wavelength by the following expression (de Broglie wavelength):

$$\lambda = \frac{h}{p}. \tag{11-3}$$

Thus it was established that "classical" waves and particles have a dual nature: the former manifest some properties of particles, corpuscles (emission and absorption of quanta), and the latter has some properties of waves (diffraction). In order to correlate both viewpoints, quantum physics hypothesizes that "particles" as well as "waves" have the nature of wave packets bounded in space and time. This is explained in Figure 21 which

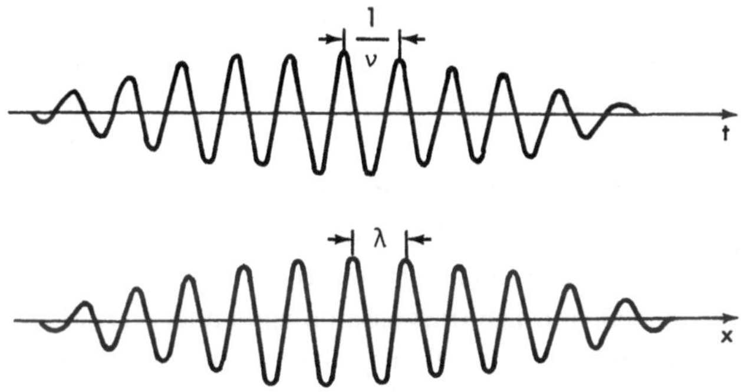

Fig. 21

shows an "instantaneous photograph" of a wave packet propagating along the x-axis and the time dependence of a wave packet at a fixed point in space. The wave packet is similar to a radar signal or any other oscillating pulse propagating in space with a finite velocity; this analogy, familar to radio engineers, makes it possible to clarify some of the principles of quantum physics. For example, let us consider the transmission of a pulse signal in a waveguide and the radiation of the pulse by a radar antenna.

Since it is desirable to match components (such as the antenna to the transmission line, etc.), the standing-wave ratio can be measured continuously, assuming that the properties of the pulse signal are similar to that of the carrier frequency. Moreover, interference between incident and reflected waves, signal phase velocity (exceeding the velocity of light), and other effects, which are characteristic only of wave phenomena, are encountered here. In measuring the antenna pattern, one actually investigates the diffraction of a continuous wave signal, again disregarding its pulse structure. If, on the other hand, the interest is in the signal delay along the transmission line, the signal behaves similar to a "particle" of finite dimensions. A pulse propagates along a transmission line at a rate lower than the velocity of light, and it is possible to measure the pulse propagation time as a whole.

With the use of a wide-band detector, it is found that energy is transmitted along the waveguide in pulses (packets) bounded in time. Also, coming closer to the conditions of known experiments, let it be imagined that it is possible to measure the radiation pressure on the waveguide wall. The mechanical effects experienced by the wall will be similar to the effects of particles impacting with a certain mass and following one another with the repetition of the pulse frequency.

Thus, pulse signals also have a dual nature, *i.e.*, wave and corpuscular. Of course, it is easy to reconcile these "contradictory" viewpoints; to do this, it suffices to represent the pulse signal as the superposition of oscillations (Fourier integral). For a small spectrum width, all "wave" phenomena practically occur as for the case of continuous wave oscillation (at the carrier frequency), and this fact does not contradict the discrete (in time and in space) structure of the signal.

Thus quantum physics and pulse technology frequently deal with similar concepts, *i.e.*, wave packets which are similar (of course, only in the formal sense) to pulsed oscillatory signals. However, it should be noted that a pulse signal can be divided into arbitrary components and each of these components can be directed along a different channel: one of the components is transmitted, another is absorbed, still another is reflected, etc. The indivisibility of signals is unknown in pulse technology, and this fact substantially distinguishes it from quantum physics.

Another fundamental characteristic of quantum mechanics is its statistical nature. Any motion of an elementary particle occurs under the condition of multiple interactions with other particles. The situation is also similar for the motion of ordinary bodies (*e.g.*, collisions between air molecules or the pressure of light caused by the "impact" of photons), but as a rule, it can be neglected when investigating macroscopic motions. Even if it is desirable to consider these effects, it suffices to consider only mean values, such as air resistance, and to ignore the details concerning their statistical nature. Therefore, classical physics can operate in the majority of cases with such deterministic concepts as velocity, coordinates of a body, force, etc. The situation is different in quantum physics. Even if one could imagine an isolated particle moving without interacting with other particles, its behavior would not be of interest. To establish the presence of a particle and to determine its position, it is necessary to illuminate this particle with light, *i.e.*, to subject it to the action of a beam of particles and to observe the reflection from or absorption by the particle being investigated. Any interaction of this kind is associated with an energy exchange between the particle under investigation and the particle which is the tool. Since the energy can only be transferred in finite units (quanta), any observation or interaction introduces a finite perturbation into the phenomenon under investigation. Since the energy associated with the motion of elementary particles is small with respect to the energy of one quantum, individual interactions cannot, in principle, be neglected. Averaging, *i.e.*, considering the mean-statistical effect alone, is justifiable only for investigations on the macroscopic scale. Thus, when such an averaging begins, the requirement for quantum physical considerations ceases to exist, and we go over to the macro-world of classical physics.

It is asserted in quantum theory that the statistical nature is an inherent characteristic of any motion. Therefore, quantum physics does not investigate the motion parameters of a single particle such as its coordinates, velocity, etc., but considers the statistical relationships which define the probability of measuring various values of these parameters.

Let us now return to the wave packet which from the present day viewpoint represents any type of elementary particle. Experiments have confirmed that the wave properties are characteristic of all particles and thus, are universal for all matter. But what is actually oscillating in a particle? What substance causes changes as shown on the graph in Figure 21? It could be assumed that for a photon the question involves the definition of intensity for electric or magnetic fields. However, classical electrodynamics defines field intensity for the macroscopic phenomena. From the quantum theory viewpoint, the field intensities must be interpreted as a result of the superposition of many photons. The use of this concept to establish the nature of a single photon requires some special substantiation. And again, what is it in this case which oscillates in an electron? Is it an electrical charge? Also, how can the wave properties of neutrons be explained?

In quantum theory, the Schrödinger function Ψ is oscillatory. This function characterizes the motion of a particle and thus depends on the conditions under which the motion is taking place. The wave function is not directly observed during measurement (at least up to the present time). Its role is manifested through other quantities measured experimentally. This is why the wave function does not give a clear idea of what actually happens "inside" the particle. One of the originators of quantum physics, Dirac, ([24], p. 26) wrote the following on this subject: "The main object of the physical science is not the provision of pictures, but is formulation of the laws governing phenomena and the application of these laws to the discovery of new phenomena. If a picture exists, so much the better; but whether a picture exists or not is a matter of only secondary importance. In the case of atomic phenomena, no picture can be expected to exist in the usual sense of the word picture, by which is meant a model functioning essentially on the classical lines."

Such an approach is also not alien in classical physics. As a matter of fact, what is the intensity of the electromagnetic field? It is not difficult to understand that this quantity has not been measured directly in any physical experiment. Our devices measure the field energy which is proportional to the square of the absolute value of the field strength, and the strength itself is determined subsequently by calculation.* Only the habit of many years makes this concept clear to physicists and engineers.

The wave function Ψ is a complex function which depends on the spatial coordinates and time:

$$\Psi(x, y, z, t) = |\Psi|\, e^{j\varphi};$$

its phase component characterizes the wave properties of the particles. In the simplest case, this phase component represents a traveling wave propagating along its coordinate axis:

$$e^{j(kx-\omega t)},$$

where $k=2\pi/\lambda$, and $\lambda=h/p$ is the de Broglie wavelength.

The wave function is directly associated with the probabilistic nature of microparticle motion. *The square of the absolute value of the wave function, $|\Psi(x, y, z, t)|^2$, is proportional, at the chosen instant of time t, to the probability (more accurately, to the probability density) that the particle will be detected at the point* (x, y, z) *in space.* Denoting $dP(M)$ as the probability of detecting the particle in the incremental volume dv near the point M, we have

$$dP(M) = \frac{1}{K}\, |\Psi|^2\, dv\,. \tag{11–4}$$

The proportionality factor K is determined from normalizing the function so that the probability of detecting the particle somewhere in unbounded space is equal to unity. For a finite volume, we obtain

$$P_v = \frac{\int_v |\Psi|^2\, dv}{\int |\Psi|^2\, dv}, \tag{11–5}$$

where the denominator is integrated over the entire unbounded space.

* In this connection, it should be noted that phase measuring devices (*e.g.*, interferometers) in reality do not measure the phase but only oscillation amplitudes which are a function of the phase relationship of the interacting waves.

Such a definition of the amplitude component of the wave equation can be interpreted as a generalization of the electrodynamics concept. The square of the magnetic field amplitude in an incremental volume is proportional to the energy of the field there. However, energy is transferred by individual quanta and, consequently, for macroscopic averaging, the energy is proportional to the mean number of photons entering the incremental volume, *i.e.*, proportional to the probability of finding an individual photon in this volume (which is sufficiently large). Thus it can be assumed that for photons under macroscopic averaging, the amplitude of the wave function converts into the amplitude of the field intensity.*

Physical laws are postulated on observations; then the conclusions reached are verified experimentally. It can be assumed that this was the way of postulating the basic equation of quantum mechanics, the Schrödinger's equation, which is the wave function depending on the space coordinates and time. This equation makes it possible to find the wave function (consequently, the probability of finding the particle) for every point in space. Similar to classical mechanics from which the velocity, energy, and other physical quantities can be found to form the laws of motion, quantum mechanics provides means for determining the probability distribution of these quantities from the wave function Ψ which is a solution of the Schrödinger equation. For the purposes of this discussion, it is not necessary to go into the calculation of these quantities. However, it is important to note that it is not possible to calculate the exact value of velocity (momentum), energy, or other parameters for an elementary particle, but only the probability of the various values.

The function Ψ contained in the Schrödinger equation is a wave function in space coordinates which defines the probability of a particle being in a given position. There are also other wave functions: Ψ_p is the wave function in momentum coordinates which defines the probability of a particle having the momentum p; Ψ_E is the wave function in energy coordinates defining the probability of the particle having the energy E; and others. For every physical quantity ϱ observed experimentally, it is possible to find a wave function Ψ_ρ such that the square of its amplitude is proportional to the probability of detecting the particle in the interval ϱ to $\varrho + d\varrho$:

$$dP(\varrho) = \frac{1}{K} |\Psi_\varrho|^2 \, d\varrho \,. \tag{11-6}$$

All wave functions are related to each other and to the "fundamental" Ψ by certain differentials, integrals, and other transforms, *i.e.*, operators.

* This correspondence is not completely accurate, but the detailed features of photon wave functions is not important to the discussion here.

[§12] UNCERTAINTY PRINCIPLE OF QUANTUM PHYSICS

We shall now return to the initial subject of the previous section, *i.e.*, the Heisenberg uncertainty principle.

Among the various physical parameters, there are sets, the wave functions of which are Fourier transform pairs. For the one-dimensional case, this means that the two wave functions, Ψ_q and Ψ_p, corresponding to two parameters, q and p, being observed, are related to each other by the following equations:

$$\Psi_p = \frac{1}{\sqrt{h}} \int_{-\infty}^{\infty} \Psi_q \, e^{-j2\pi \frac{qp}{h}} \, dq, \tag{12–1a}$$

$$\Psi_q = \frac{1}{\sqrt{h}} \int_{-\infty}^{\infty} \Psi_p \, e^{j2\pi \frac{qp}{h}} \, dp. \tag{12–1b}$$

It will now be shown that the deviation of the two variables p and q from their mean values $\bar{p}$ and $\bar{q}$ are related. If, under the experimental conditions (*e.g.*, a special setup for measuring one of the parameters), the measurement error δp is small, the measurement error of the conjugate variable δq will inevitably be larger. For the standard deviations of the measurements, the uncertainty relationship is

$$\sigma_p \sigma_q \geqslant \frac{h}{4\pi} = \frac{\hbar}{2}. \tag{12–2}$$

This establishes the accuracy limit when p and q are measured simultaneously.

In evaluating the coefficient K for equations like (11–6), it is assumed, first of all, that the function Ψ_p is normalized to the following condition:

$$\int_{-\infty}^{\infty} |\Psi_p|^2 \, dp = 1.$$

It follows from Parseval's theorem and (12–1) that in this case the conjugate function Ψ_q is also normalized in the same manner:

$$\int_{-\infty}^{\infty} |\Psi_q|^2 \, dq = \int_{-\infty}^{\infty} |\Psi_p|^2 \, dp = 1.$$

From the definition of the wave function, the mean value $\bar{p}$ and its standard deviation σ_p are defined by the following equations:

$$\bar{p} = \int_{-\infty}^{\infty} p\,|\Psi_p|^2\,dp\,,$$

$$\sigma_p^2 = \int_{-\infty}^{\infty} (p - \bar{p})^2\,|\Psi_p|^2\,dp\,.$$

There are similar equations for the variable q. It is possible to translate the origin of each variable by the amount of its mean, *i.e.*, to force $\bar{p} = \bar{q} = 0$.

Let us now turn to proving (12–2). We start by forming the function

$$f(x) = \int_{-\infty}^{\infty} \left| xp\Psi_p - \frac{d\Psi_p}{dp} \right|^2 dp\,. \tag{12–3}$$

Due to the absolute value signs under the integral, $f(x) \geqslant 0$. Reducing the integrand in this equation to a product of conjugate factors and performing the multiplication, we obtain

$$f(x) = ax^2 + bx + c\,, \tag{12–4}$$

where

$$a = \int_{-\infty}^{\infty} p^2\,|\Psi_p|^2\,dp\,,$$

$$b = \int_{-\infty}^{\infty} p\left(\Psi_p \frac{d\Psi_p^*}{dp} + \Psi_p^* \frac{d\Psi_p}{dp}\right) dp$$

$$= \int_{-\infty}^{\infty} p(\Psi_p \cdot \Psi_p^*)'\,dp\,,$$

$$c = \int_{-\infty}^{\infty} \left|\frac{d\Psi_p}{dp}\right|^2 dp\,.$$

For all real values of x, the quadratic trinomial of (12–4) is positive, and thus it cannot have real roots. Therefore,

$$b^2 - 4ac \leqslant 0\,. \tag{12–5}$$

Let us now find the values of the coefficients a, b, and c. From the definition of the standard deviation,

$$a = \int_{-\infty}^{\infty} p^2 |\Psi_p|^2 \, dp = \sigma_p^2 .$$

Integrating by parts for the expression defining the coefficient b, we have

$$b = p|\Psi_p|^2 \Big|_{-\infty}^{\infty} - \int_{-\infty}^{\infty} |\Psi_p|^2 \, dp .$$

Since the integral defining the coefficient a converges to a finite value, the first term in the above equation must vanish; therefore, considering the normalization of the function Ψ_p, we find

$$b = -\int_{-\infty}^{\infty} |\Psi_p|^2 \, dp = -1 .$$

To determine the coefficient c, we first differentiate (12–1a) with respect to p:

$$\frac{d\Psi_p}{dp} = -\frac{4\pi j}{h^{3/2}} \int_{-\infty}^{\infty} q\Psi_q \, e^{-j2\pi\frac{qp}{h}} \, dq .$$

Using Parseval's theorem, we find

$$c = \int_{-\infty}^{\infty} \left|\frac{d\Psi_p}{dp}\right|^2 dp = \frac{4\pi^2}{h^2} \int_{-\infty}^{\infty} q^2 |\Psi_q|^2 \, dq = \frac{4\pi^2}{h^2} \sigma_q^2 .$$

Substituting the values of these coefficients into (12–5), we obtain

$$\frac{16\pi^2}{h^2} \sigma_p^2 \sigma_q^2 \geqslant 1 ,$$

which is equivalent to the uncertainty relationship (12–2).

After some consideration, it is realized that nothing unexpected has been obtained. Equations (12–1) show that the wave functions Ψ_p and Ψ_q are related to each other in the same manner as the pulse signal $s(t)$ is related to its spectrum $S(\omega)$. The variables p and q have the same mathematical meaning as the time t and the frequency ω (ignoring the constant factor $2\pi/h$). Therefore, the uncertainty relationship (12–2) is nothing more than a rigorous formulation of the well-known principle that the product of the signal duration by its spectrum width cannot be arbitrarily decreased and that this product is always greater than a given constant:

$$\Delta t \cdot \Delta\omega \geqslant \mu . \tag{12-6}$$

It is frequently assumed that $\mu \approx 1$; however, this constant is of course dependent on the definition of pulse duration and spectrum width. Thus, for different definitions of duration and width, coupled with different pulse forms, various values of μ are possible; however, for any "practical" set of definitions, the constant μ is the order of unity. This signal characteristic (more accurately, this property of Fourier conjugate functions) has been used in §4. The above derivation contains a definition of duration and pulse width suitable for any signals. The duration of signal $s(t)$ is defined as follows. The origin of the time scale is placed at the "center" of the pulse, so that

$$\int_{-\infty}^{\infty} t\,|s(t)|^2\,dt = 0,$$

and the signal function is normalized so that the energy $E=1$, *i.e.*,

$$\int_{-\infty}^{\infty} |s(t)|^2\,dt = \frac{1}{2\pi}\int_{-\infty}^{\infty} |S(\omega)|^2\,d\omega = 1 .$$

Then the pulse duration T is defined by the expression

$$T^2 = \sigma_t^2 = \int_{-\infty}^{\infty} t^2\,|s(t)|^2\,dt .$$

Similarly, the spectrum width F is defined by the integral

$$4\pi^2 F^2 = \sigma_\omega^2 = \frac{1}{2\pi}\int_{-\infty}^{\infty} \omega^2\,|S(\omega)|^2\,d\omega,$$

with the additional condition which defines the origin of the frequency scale,

$$\int_{-\infty}^{\infty} \omega\,|S(\omega)|^2\,d\omega = 0 .$$

It follows from the previous derivation that for pulse duration and spectrum width defined in such a manner, the condition

$$\sigma_\omega \sigma_t \geqslant \tfrac{1}{2}$$

or

$$FT \geqslant \frac{1}{4\pi} \tag{12–7}$$

is valid for any signal form. This relationship reflects the general property of functions which are Fourier transform pairs. Eq. (12–7) can be termed as the "*indirect relationship*". However, in order to convert (12–7) into the "uncertainty relationship" which expresses the ultimate potentialities of any physical measurement, additional assumptions are needed. As a matter of fact, these additional conditions are the same as those formulated (more precisely, postulated) in quantum theory on the basis of atomic scale experimental data. The Heisenberg uncertainty principle is a direct consequence of the basic postulates of quantum mechanics and has been confirmed by experimentation.

In addition to the coordinate and momentum of a particle, quantum mechanics also provides the uncertainty relationship between the time t that a particle appears at a chosen point in space, and its energy E at this point. For the standard deviations of the measurement, σ_t and σ_E, we have the relationship

$$\sigma_E \sigma_t \geqslant \frac{h}{4\pi}\,. \tag{12–8}$$

Since the energy E is related to the frequency ν by

$$E = h\nu\,,$$

the standard deviation of the frequency can be inserted into (12–8) and we obtain

$$\sigma_\nu \sigma_t \geqslant \frac{1}{4\pi}\,. \tag{12–9}$$

It appears that this equation is the same as (12–7) (the frequency notation ω is replaced by $2\pi\nu$, as is customarily done in quantum physics), but it has an entirely different meaning. Eq. (12–7) is not a relationship between measurement errors; it merely expresses the interrelationship between the pulse duration and its spectrum width. Eq. (12–8) contains an "identification" with quantum theory, *i.e.*, Planck's constant h. Without stating it explicitly, this equation contains the quantization of energy, the stepwise character of physical processes, and the entire probabilistic interpretation of the phenomena on the atomic scale. In quantum physics, this equation reflects the fundamental uncertainty in measuring t and E. Equation (12–9) has the same meaning; in this equation, ν is understood not as a frequency of an arbitrary macroscopic oscillation, but as the frequency (more accurately, a component of the frequency spectrum) of the wave packet.

[§13] OTHER FORMS OF THE UNCERTAINTY RELATIONSHIP

While the relationships of the previous section are the best known ones, they are by no means the only interpretations of the uncertainty principle. The equations of Heisenberg and Weyl characterize the measurement dispersion (or standard deviation) as a value of the uncertainty of the measurement. The relationship between the signal duration and its spectrum width defined in (12–7) is a similar expression.

Of course, such definitions of the signal, spectrum, and their distributions are not unique. For example, in §4 we have used another definition of the signal duration and the spectrum width. In this connection, there are also other uncertainty principles employing different definitions of the variables. We shall quote here the results of some other authors.*

The following definition of the duration of the function $s(t)$ over an infinite interval is used in the work of Zakai [77]:

$$T_u = \left[\frac{\left(\int\limits_{-\infty}^{\infty} |s(t)|^2 \, dt \right)^{1/2}}{\left(\int\limits_{-\infty}^{\infty} |s(t)|^u \, dt \right)^{1/u}} \right]^{\frac{2u}{u-2}}. \tag{13–1}$$

It is assumed here that both integrals converge, and that the parameter u can have any arbitrary positive value. Of course, the duration T_u depends on this parameter. For the special case of $u=0$, $u=2$, and $u=\infty$, the duration is understood to be limit of T_u as u approaches these values. These cases are used in numerous works.

It can be shown that for $u=0$, the quantity T_0 is a measure of the length of t for which $s(t) \neq 0$. T_0 is the total duration of a pulse with finite length.

The case of $u=\infty$ reduces to the definition of duration that was used in §4. In fact, it follows from the relationship

$$\lim_{u \to \infty} \left(\int\limits_{-\infty}^{\infty} |s(t)|^u \, dt \right)^{1/u} = \max |s(t)|$$

* The well-known refinements of the Heisenberg and Weyl uncertainty relationship are based on the standard deviation of the uncertainty and are not considered in the following discussion. Such refinements can be found, for example, in the books of Kharkevich ([68], p. 82) and Helstrom ([70], p. 21).

that

$$T_\infty = \frac{\int_{-\infty}^{\infty} |s(t)|^2 \, dt}{(\max s(t))^2}.$$

The case of $u=2$ requires a more detailed explanation. Denoting the integral in the numerator of (13–1) by E, we have*

$$T_{2+\delta} = \left[\frac{E^{1/2}}{\left(\int_{-\infty}^{\infty} |s(t)|^{2+\delta} \, dt \right)^{\frac{1}{2+\delta}}} \right]^{\frac{4+2\delta}{\delta}}.$$

Furthermore, from the approximation

$$|s(t)|^{2+\delta} = |s(t)|^2 \, e^{\delta \ln |s(t)|} \approx |s(t)|^2 (1 + \delta \ln |s(t)|)$$

we obtain

$$T_2 = \lim_{\delta \to 0} \frac{E}{\left(1 - \delta \dfrac{\mathscr{E}_t}{2E}\right)^{\frac{2}{\delta}}} = E \, e^{\frac{\mathscr{E}_t}{E}}, \tag{13–2}$$

where

$$\mathscr{E}_t = - \int_{-\infty}^{\infty} |s(t)|^2 \ln |s(t)|^2 \, dt. \tag{13–3}$$

In quantum physics, $s(t)$ is the wave function and $|s(t)|^2$ is the probability density of the variable t being observed. Therefore, the integral (13–3) is the *entropy* parameter frequently used in information theory and thermodynamics as a measure of uncertainty (see §22). Consequently, the definition (13–1) includes, as a special case, the entropy measure of uncertainty.

If $s(t)$ (a signal) is a function of time, the spectrum width can be defined in a similar manner:

$$F_v = \frac{1}{2\pi} \left[\frac{\left(\int_{-\infty}^{\infty} |S(\omega)|^2 \, d\omega \right)^{1/2}}{\left(\int_{-\infty}^{\infty} |S(\omega)|^v \, d\omega \right)^{1/v}} \right]^{\frac{2v}{v-2}}. \tag{13–4}$$

* Here, $s(t)$ and E are dimensionless; in general, $E \neq 1$.

The uncertainty relationship proved by Zakai has the form

$$T_u F_v \geqslant 1;$$

moreover, this relationship is valid for any functions $s(t)$ and $S(\omega)$ that are Fourier transform pairs, provided the parameters u and v satisfy the condition

$$\frac{1}{u} + \frac{1}{v} \geqslant 1.$$

A more complete result for the entropy measure of uncertainty was obtained by Leipnik ([45], pt. 1). He showed that for the normalized condition of $E=1$, the entropies $\mathscr{E}_f$ and $\mathscr{E}_t$ [the latter is defined similar to (13–3)] are related by*

$$\mathscr{E}_t + \mathscr{E}_f \geqslant -\ln\frac{1}{2} = 0.6931. \tag{13–5}$$

In particular, this equation reveals the additive nature of entropy; the resulting uncertainty is a function of the sum of entropies and not of their product (as in the case of the Heisenberg-Weyl formulation). However, it is interesting to note that the optimal function which ensures a minimum uncertainty is the same for both definitions of entropy; the uncertainty is estimated by the standard deviation of the measurement. The Gaussian curve $s(t)=e^{-t^2/2}$ is such a function.

The uncertainty relationships of Zakai and Leipnik "extend" over a comparatively wide class of definitions but, unfortunately, these definitions do not have physical clarity. The results of Landau and Pollack ([41], pt. II) are more complete in this sense and are presented below.

Let us denote by α^2 the ratio of the signal energy in time interval $t_0 - T/2 \leqslant t \leqslant t_0 + T/2$ to the total signal energy:

$$\alpha^2 = \frac{\int\limits_{t_0-T/2}^{t_0+T/2} |s(t)|^2\, dt}{\int\limits_{-\infty}^{\infty} |s(t)|^2\, dt}. \tag{13–6}$$

Similarly, β^2 denotes the ratio of energy contained in the frequency interval

* In quantum mechanics where conjugate wave functions are Fourier transform pairs as in (12–1), the right-hand side is replaced by log $he/2$; moreover, the base of the logarithms in the entropy inequality may be arbitrarily chosen.

$-\Omega/2 \leqslant \omega \leqslant \Omega/2$:

$$\beta^2 = \frac{\int_{-\Omega/2}^{\Omega/2} |S(\omega)|^2 \, d\omega}{\int_{-\infty}^{\infty} |S(\omega)|^2 \, d\omega}. \tag{13–7}$$

The signal duration and spectrum can be defined as the time and frequency intervals which contain the designated ratios of the total energy. Such definitions are physically justifiable and widely used by radio engineers. It was shown by Landau and Pollack that the following uncertainty relationship holds for any arbitrary signal $s(t)$ at any arbitrary time t_0:

$$\Omega T \geqslant \Phi(\alpha, \beta), \tag{13–8}$$

where Φ, which is a function of the energy ratios α^2 and β^2, is determined in the following manner. When the product ΩT is held constant, the energy ratios α^2 and β^2 are bounded, and it is possible to formulate the uncertainty relationship (13–8) by determining the upper boundaries of the energy ratios for the constant value of ΩT. These boundaries are given by the following relationships:

$$0 \leqslant \beta < 1 \quad \text{for} \quad \alpha = 0,$$
$$0 \leqslant \beta \leqslant 1 \quad \text{for} \quad 0 < \alpha < \sqrt{\lambda_0},$$
$$\cos^{-1}\alpha + \cos^{-1}\beta \geqslant \cos^{-1}\sqrt{\lambda_0} \quad \text{for} \quad \sqrt{\lambda_0} \leqslant \alpha < 1,$$
$$0 < \beta \leqslant \sqrt{\lambda_0} \quad \text{for} \quad \alpha = 1.$$

The value of λ_0 depends on the product ΩT and is the largest eigenvalue of the integral equation

$$\lambda\psi(t) = \frac{1}{\pi} \int_{-T/2}^{T/2} \psi(t') \frac{\sin \frac{\Omega}{2}(t - t')}{t - t'} \, dt'. \tag{13–9}$$

The values of the energy ratios for several ΩT products are given in Figure 22. It is clear from this graph that to concentrate 90% of the total energy ($\alpha^2 = \beta^2 = 0.9$) in the T and Ω intervals, it is necessary to satisfy the following condition:

$$\Omega T \gtrsim 8.$$

The optimal functions which result in a minimum value of ΩT for given energy ratios α^2 and β^2 (and conversely, maximum values of α^2 and β^2 for a given ΩT) are closely related to the eigenfunction $\psi_0(t)$ in (13–9) that

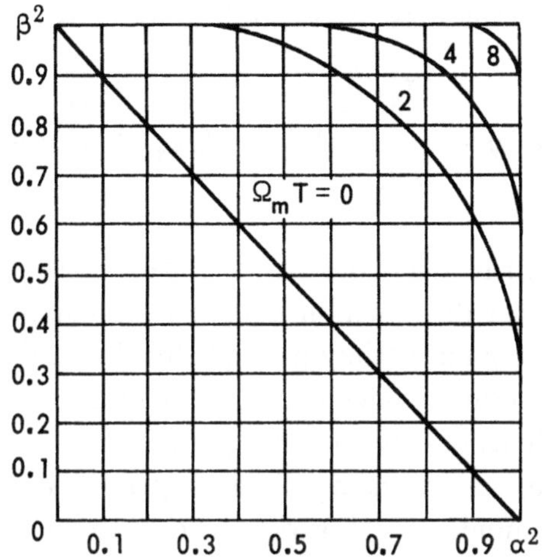

Fig. 22

corresponds to the minimum eigenvalue, λ_0. This subject is discussed in greater detail in §28 and Appendix 5.

In conclusion, we shall investigate another form of expressing uncertainty, which is close to the radar uncertainty relationship developed in §10. This form of expressing uncertainty is related to the probability density function of Wigner and is sometimes used in quantum physics (see Leipnik [45], pt. II, and also Klauder [33]).

As a rule, the probability distribution of a particle's coordinate q and its conjugate p, the momentum, can be considered independently. In this case, the wave functions Ψ_q and Ψ_p satisfy the relationships (12–1) and contain all the necessary information. But sometimes there is the need for determining the combined wave function Ψ_{qp} related to the conditional probability of detecting a particle at point q and having the momentum p. If the particle's coordinate and momentum are statistically independent, the wave functions are related by the following expression:

$$|\Psi_{qp}|^2 = |\Psi_q|^2 \, |\Psi_p|^2 . \tag{13–10}$$

However, (13–10) does not contain the phase of function Ψ_{qp}; this information can be obtained from the expression

$$\Psi_{qp} = \Psi_q^* \Psi_p \, e^{j \frac{2\pi}{h} qp} .$$

The double Fourier transform of these wave functions is

$$\varphi(q, p) = \frac{1}{h} \int_{-\infty}^{\infty} \int_{-\infty}^{\infty} \Psi_{q'p'} \, e^{j \frac{2\pi}{h} (qp' + pq')} \, dq' \, dp' .$$

Some simple transformations [see (8–4)] lead to the relationship

$$\varphi(q, p) = \frac{e^{j\frac{\pi}{h}qp}}{\sqrt{h}} \int_{-\infty}^{\infty} \Psi^*_{q'-q/2} \Psi_{q'+q/2} \, e^{j\frac{2\pi}{h}pq'} \, dq' .$$

The function $\varphi(q, p)$ is analogous to the radar ambiguity function. It can be readily shown that this function $\varphi(q, p)$ is "incompressible," and this reflects the Heisenberg uncertainty principle for the particle coordinate and its conjugate, momentum. Evidently, there exists here a formal analogy with radar. The wave functions of quantum physics are directly related to the probabilities of the corresponding physical quantities. The square of the wave function magnitude, $|\Psi_{qp}|^2$, is proportional to the probability that the particle position will be in the interval from q to $q+dq$ and that its momentum will be between p and $p+dp$. If the radar measurement uncertainties for determining a target's position and range rate would follow the same statistical laws as the microworld, the same relationships would apply. Actually, to do this, it must be assumed that the probability of measuring a signal's time delay and frequency shift in the intervals t to $t+dt$ and Ω to $\Omega+d\Omega$, respectively, has the form

$$dP(t, \Omega) = \frac{1}{E} |s(t)|^2 \, |S(\Omega)|^2 \, dt \, d\Omega .$$

Only under this additional condition would the radar ambiguity function lead to the same uncertainty principle as in quantum physics. It was shown in §8 that the magnitude of the ambiguity function $\chi(t, \Omega)$ defines the signal level at the output of an optimal receiver for various target ranges and range rates. For a constant noise level, the probability of the various range and range rate errors is a function of the signal level. However, by assuming that the ambiguity function (or the antenna pattern) is directly related to the measurement distribution, we can formulate, without a proof, important statistical properties of the measurements while not specifying the nature of the errors and the noise level. Consequently, the validity of the radar uncertainty principle cannot be asserted only on the basis of this analogy with quantum theory.

This remark pertains fully to all forms of the ambiguity function considered above. We have seen that for many "practical" functions which are Fourier transform pairs, there exist relationships which are similar to the Heisenberg-Weyl function. However, the probabilistic interpretation of such relationships can be justified only by the physical nature of the phenomenon and not by the mathematical similarity.

[§14] RESOLUTION AND SELECTIVITY OF A LINEAR MEASURING DEVICE

Let us now consider the problem of measuring frequency where, at the first glance, the radar uncertainty principle has been confirmed by physical measurement.

It is well known that the higher the frequency resolution, the longer the analysis time, *i.e.*, the time required to analyze the signal spectrum (*e.g.*, see Kharkevich [68], p. 124). If a single pulse is considered, this time does not exceed the pulse duration. In addition, if everything else is the same, the longer the pulse duration, the lower the resolution of the time of arrival measurement. Thus, the frequency and time resolutions are reciprocal, and their product is bounded by a constant:

$$\Delta\omega\,\Delta t \geqslant \mu\,, \tag{14–1}$$

which characterizes the ultimate combined resolution. There is a similar relationship for the time and frequency measurement resolutions.

Since the radar measurement of range and range rate reduces to measuring the signal arrival time and frequency, (14–1) can be interpreted as the radar uncertainty relationship. In order to refine this statement, it is necessary to use a general theory of linear measuring devices.

There are many measuring devices which are linear. A linear measuring device is defined as one in which the response of the device, $y(t)$, is related to the input signal, $x(t)$, by the linear integral transform

$$y(t) = \int_{-\infty}^{\infty} h(t-t')\,x(t')\,dt'\,. \tag{14–2}$$

Of course, this transform corresponds to a linear electrical circuit, in particular the radar receiver. In this case, the independent variable is t, $x(t)$ and $y(t)$ are the electrical signals at the input and output, respectively, and $h(t)$ is the circuit impulse response function, *i.e.*, the output for a δ-function input.

Equation (14–2) also corresponds to optical devices, *i.e.*, a telescope, microscope, etc. In this case, t is the spatial (angular) coordinate (a one-dimensional model is considered here), and $x(t)$, $y(t)$ are the brightness distributions in the planes of the object and image, respectively. The function $h(t)$ is now the diffracted image of a point source, *i.e.*, the response to a δ-function. Here the δ-function is with respect to a point in space instead of an instant of time.

Let us also consider the measurement of frequency with an ordinary tuned wavemeter. Assume a monochromatic signal is received at the wavemeter input, *i.e.*, the input signal is a δ-function along the frequency axis. By tuning the wavemeter, we pass through its response to such a δ-function. The frequency characteristic of the wavemeter is this response. If a sophisticated signal containing many frequency components is supplied to the input, the tuning of the wavemeter results in a reading composed of the superposition of the responses of the separate frequency components. This curve represents the signal's frequency spectrum obtained from a measurement with some distortions. It can be readily shown that the curve corresponds to the integral (14–2) where t is the frequency, $h(t)$ the frequency characteristic of the wavemeter,* and $x(t)$, $y(t)$ are the true spectrum and the measured spectrum with distortion, respectively. Thus, (14–2) is also applicable to spectrum analyzers.

The function $h(t)$ fully characterizes the properties of a measuring device and is frequently called the *equipment function*. However, $h(t)$ is a complicated complex function and its use is not convenient for comparing various devices. Therefore, a simpler function defining the device's characteristics is frequently used for comparison.

Resolution is an important parameter of a measurement device; it defines to what extent the "fine structure" of the input signal can be analyzed. The classical definition of resolution is obtained from the following considerations. Let the input signal consist of two δ-functions separated from one another by the interval Δt. This can be two electrical pulses with Δt being the time interval between them, or two point sources of light where Δt is the space interval, or two monochromatic signals with Δt being their frequency difference, etc.

The response of a device to such an input is in the form of two spikes of finite width, each of which is the device's impulse response function $h(t)$ (see Fig. 23a). By decreasing the separation Δt, these spikes come closer to each other and finally blend together. The resulting response takes the form shown in Figure 23c. It is customary to assume that the device cannot resolve such signals. If, however, the resulting response has a double-humped form (see Fig. 23b), it is assumed that the signal can be resolved.

The classical definition of resolution is based on the qualitative discrimination between signals. *According to Rayleigh, the resolution of a device is the minimum interval Δt for which the resulting response still has the form of a double-humped curve.* More accurately, the interval corresponds to the "critical" case where the double-humped curve just transforms into a single-humped curve.

* Strictly speaking, $h(t)$ is the dynamic frequency characteristic which is a function of the speed with which the wavemeter is tuned (dynamic transmission coefficient). If the tuning is sufficiently slow, $h(t)$ becomes the ordinary frequency characteristic.

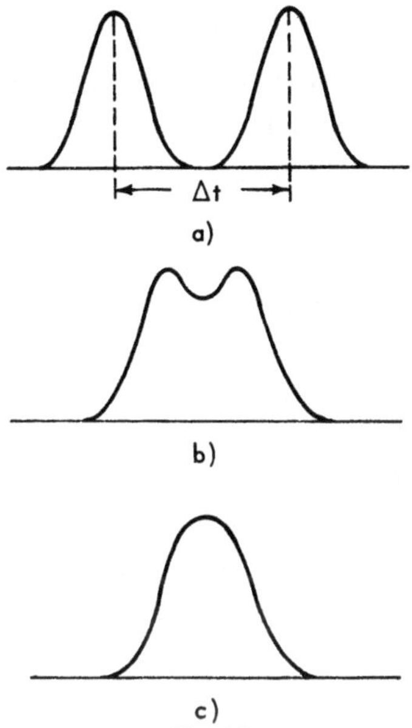

Fig. 23

It is easy to notice that for a Gaussian-shaped response curve $h(t)$, the resolution of the device defined in the above manner depends directly on the width of the curve. If, for example, $h(t)$ is a Gaussian function, the response for two δ-functions has the form

$$y(t) = h\left(t - \frac{\Delta t}{2}\right) + h\left(t + \frac{\Delta t}{2}\right) = \mathrm{e}^{-\frac{\left(t - \frac{\Delta t}{2}\right)^2}{\tau^2}} + \mathrm{e}^{-\frac{\left(t + \frac{\Delta t}{2}\right)^2}{\tau^2}}.$$

The resolution is defined by the value of Δt for which the second derivative $y''(t)$ vanishes at the point $t=0$. This follows from the obvious fact that for resolved signals, the response $y(t)$ has a minimum at $t=0$, and for non-resolved signals, the response has a maximum at this point. Performing the necessary mathematical operations, we have

$$y''(0) = -\frac{4}{\tau^2}\,\mathrm{e}^{-\frac{\Delta t^2}{4\tau^2}}\left(1 - \frac{\Delta t^2}{2\tau^2}\right),$$

and the condition for $y''(0)=0$ defines the minimum resolution interval:

$$\Delta t_{\min} = \sqrt{2}\tau\,.$$

Thus, for an electrical circuit, the resolution depends directly on the duration of the impulse response, and for a wavemeter, on the width of the frequency characteristic, etc.

In the following, it will be shown that it is this Rayleigh definition of resolution which leads to the uncertainty relationship (14–1). However, the arbitrariness of this definition must be emphasized. In the first place, it is implied that the signal has a particular form, *i.e.*, two δ-functions separated by some interval. It is not clear how to apply the above concept to signals with different forms. Secondly (and this is the more important reason), the resolution criterion was arbitrarily chosen. It can only be said that such a criterion (double- or single-humped curves) points to a qualitative, easily detectable, distinction between responses. In principle, nothing has been said here that it is impossible to resolve two signals when the interval between them is smaller than $\Delta t_{\min}$.* No matter how small the interval Δt is made, the resulting response differs, nevertheless, from that obtained for a single impulse and, consequently, it is possible, in principle, to distinguish whether one or two signals are present at the input. It is natural, therefore, that the classical definition of resolution is not consistent with experimental results. In particular, the photometric methods perfected for investigating telescopic images make it possible to exceed substantially the calculated resolution (Garelik, [15]).

Another important characteristic of a measuring device is its selectivity with respect to the controlled parameter (independent variable). Let the input signal, again being a δ-function, arrive to the input at time t_0 (in this example an electrical circuit is being considered so that the controlled parameter t denotes time). The response to such a signal has the form

$$y(t) = h(t - t_0).$$

If the time of the signal's arrival, t_0, is changed, the response will shift in time, and there will be a lower level of output signal $y(t)$ where previously there was a maximum. The selectivity of a device is characterized by the width Δt between the 50% and maximum power levels. Obviously, the selectivity is only the half width of the 0.7 voltage levels. It is customary to assume that such a change in output level can easily be detected and, consequently, the selectivity characterizes, to a certain degree, the accuracy of measuring the controlled parameter t. This concept has the same basic shortcoming as the definition of uncertainty. The difference in levels for making the measurement is arbitrarily chosen. However, by thoroughly investigating the output response, it is possible to detect arbitrarily small, finite signal shifts.

In principle, both of the above-mentioned definitions are unsound. The resolution and accuracy cannot be estimated correctly without considering the effects of random interference. In any kind of measurement, only the

* It should be noted that the initial definition given by Rayleigh for optical devices differs somewhat from that stated above, but they both are based on the conditional resolution criterion (Gorelik [14], pp. 471, 521).

fluctuations superimposed on the useful signal which are not removable, can cause errors. The statistical estimate of resolution and accuracy will be given in the next chapter. Here we will show that only in the absence of random distortions is it possible to obtain an arbitrarily high resolution and selectivity from a linear measuring device. Let us assume that for an input signal $x(t)$, a response $y(t)$ is obtained at the output of the device. If the impulse response function of the device, $h(t)$, is known, it is possible to reconstruct the input signal from the measured output. Then, by the same token, maximum information will be obtained from the measurement. Furthermore, it is always possible to answer "specific" questions: how many separate signals are at the input; what are the intervals between these signals; etc.

The problem of restoring the input signal reduces to solving the integral equation (14–2) with respect to $x(t)$. It is not difficult to obtain the mathematical solution of this equation. Denoting $X(\omega)$, $Y(\omega)$, and $H(\omega)$ as the spectra of $x(t)$, $y(t)$, and $h(t)$, respectively, and using the convolution theorem, we find*

$$Y(\omega) = H(\omega)\, X(\omega).$$

Thus,

$$x(t) = \frac{1}{2\pi} \int_{-\infty}^{\infty} X(\omega)\, e^{j\omega t}\, d\omega = \frac{1}{2\pi} \int_{-\infty}^{\infty} \frac{Y(\omega)}{H(\omega)}\, e^{j\omega t}\, d\omega . \qquad (14\text{–}3)$$

It is this integral that gives the requires solution, *i.e.*, computes the unknown input signal, $x(t)$, from the measured response, $y(t)$, and the equipment impulse response function, $h(t)$. It is somewhat more difficult to clarify under what conditions this solution is unique. What are the requirements to determine uniquely the input signal from the measured response? It has been proved for this purpose that it is sufficient to impose some constraints on the class of input signals. In particular, the following theorem specifies a class of signals where this holds (Khurgin & Yakovlev [71], p. 184).

For all input signals, $x(t)$, having finite duration [i.e., $x(t)$ must differ from zero only in the interval $a<t<b$, where a and b may be different for different $x(t)$], and having finite energy [i.e., the quadrature components of $x(t)$ must be integrable], there is one signal function computed from (14–3) which yields the unique solution of (14–2). And what is more, if the measured response, $y(t)$, is determined with some errors not exceeding a maximum of ε at every point, then the input signal calculated from (14–3) will have a maximum error

* It should be noted that the ω in these equations does not necessarily stand for frequency. For example, in the case of a spectrum analyzer, t represents frequency and ω represents time. It is only important that t and ω be the variables of conjugate Fourier transform pairs.

of η which depends on ε. Also, the smaller the measurement errors, the smaller becomes the quantity η ($\eta \to 0$ as $\varepsilon \to 0$).

The assumption that the input signals have a bounded duration has no practical effect on the generality of the theorem. As a rule, it can be assumed that the input signal to a receiver is bounded in time, the object under investigation has finite dimensions, the spectrum of the signal is finite, etc. Also, a signal consisting of two δ-functions has a bounded separation between these two functions, Δt. The above theorem shows that by accurately measuring the output, it is possible, in the majority of cases, to determine precisely the input signal.

Thus, for an input signal with sufficiently high energy, where the fluctuating noise and other interferences can be neglected, there is no limit to the measurement resolution. By appropriately processing the output signal, it is possible to determine precisely any parameters of the input signal and, consequently, to recover all possible information from the measurement.

[§15] RESOLUTION OF SPECTRAL ANALYSIS

Let us consider in greater detail the frequency measurement which leads to the uncertainty relationship (14–1). First of all, it is necessary to consider some of the characteristic features of the physical spectrum analysis.

There is no analyzer in existence today capable of determining the mathematical spectrum of a signal, *i.e.*, to calculate the spectral function

$$S(\omega) = \int_{-\infty}^{\infty} s(t)\, e^{-j\omega t}\, dt. \tag{15–1}$$

In making a measurement, the analyzer is turned on initially at the time $t = 0$ and turned off at $t = T$. The result of the analysis does not depend on the signal behavior prior to turning on the analyzer or after it is turned off. Therefore, the best possible result obtained from an "ideal" analyzer is

$$S_T(\omega) = \int_{0}^{T} s(t)\, e^{-j\omega t}\, dt, \tag{15–2}$$

which is usually called the present spectrum of the signal $s(t)$. For example, let the signal under investigation be infinitely long and monochromatic:

$$s(t) = e^{j\omega_0 t}.$$

Mathematically, the spectrum of such a signal is a δ-function:

$$S(\omega) = 2\pi\delta(\omega - \omega_0).$$

This signal is received by the analyzer for a finite time interval; "to the analyzer" the input signal is a pulse having a rectangular envelope. The spectrum obtained from the measurement is

$$S_T(\omega) = \int_0^T e^{-j(\omega-\omega_0)t}\,dt = 2\frac{\sin(\omega-\omega_0)\dfrac{T}{2}}{\omega-\omega_0}\,e^{-j(\omega-\omega_0)\frac{T}{2}}, \tag{15–3}$$

and this spectrum has a finite width the order of $1/T$. Thus, instead of an infinitely narrow spectrum, we measure a spectrum of finite width which depends on the duration of analysis, T.

If the input signal is a sum of two sinusoids (two δ-functions on the frequency axis), the measured spectrum is the superposition of two responses of the type expressed by (15–3). For a small difference in frequency, the envelope of the resulting spectrum has the form of a single-humped curve. It can be readily shown that the resolution of an analyzer (using the classical Rayleigh definition) is inversely proportional to the analysis time (see Kharkevich [68], pp. 124–137). In the more general case, where the signal under investigation has any arbitrary spectrum $S(\omega)$, the "ideal" analyzer turned on for a finite time duration T will display the spectral function:*

$$S_T(\omega) = \int_{-\infty}^{\infty} H(\omega - \omega')\,S(\omega')\,d\omega'. \tag{15–4}$$

From (15–3), the spectrum of the analyzer's impulse response function, $H(\omega)$, has the form

$$H(\omega) = \frac{1}{\pi}\frac{\sin\omega\dfrac{T}{2}}{\omega}\,e^{-j\omega\frac{T}{2}}.$$

The general theory of linear measuring devices considered in the previous section can be directly applied to this case. It follows from §14 that under very general assumptions, the signal spectrum, $S(\omega)$, can be reconstructed from the "measured output" of the analyzer, $S_T(\omega)$. Therefore, even a short analysis time does not lead, in principle, to an unconditional constraint of the resolution. There is no such constraint if the signal under investigation has a finite frequency band and its level is high enough to exclude the effect of fluctuations.

* Equation (15–4) is identical to (15–2) by virtue of the convolution theorem.

To explain this result, let us once again consider a signal in the form of two sinusoids:

$$s(t) = e^{j\omega_1 t} + e^{j\omega_2 t}, \quad -\infty < t < \infty .$$

Such a signal has a finite frequency band and is sampled by the analyzer over the finite time interval $0<t<T$. Thus, the input signal to the analyzer can be written as follows:

$$s(t) = a(t)\,[e^{j\omega_1 t} + e^{j\omega_2 t}],$$

where $a(t)$, the pulse envelope, is rectangular and has the duration T.

With the signal $s(t)$ at the input, a response is formed at the analyzer's output, and from this output the input signal can be reconstructed. In the strict sense, this follows from §14. However, the fact that signal reconstruction is possible can be confirmed in a simpler way. A wideband circuit which does not distort the signal noticeably can be used as the "spectrum analyzer". For example, an oscilloscope with a sufficiently small time constant ($\tau<T$) displays undistorted oscillations over the finite time interval $\tau \leqslant t \leqslant T$. It will be shown subsequently that over a finite region (within the interval of observation) the signal under investigation has the form

$$s(t) = e^{j\omega_1 t} + e^{j\omega_2 t} . \tag{15–5}$$

Of course, nothing is known about the signal outside the observation time interval; its spectrum is also unknown, but what is important here is that the spectrum has a finite frequency band. By virtue of this condition, the signal $s(t)$ is an analytical function of time and if it is known for any finite interval, it is uniquely defined over the entire range $-\infty<t<\infty$. In particular, the signal $s(t)$ cannot be different from (15–5) over the entire time axis. By determining the signal over the entire interval, we can establish that its spectral density differs from zero only at the frequencies $\omega=\omega_1$ and $\omega=\omega_2$, *i.e.*, it is possible to separate the two frequencies no matter how close they are to each other for the short duration of the analysis, T.

From the previous discussion on the requirements for arbitrary high resolution, it is necessary to satisfy two conditions. First, the signal level must be high enough so that the superimposed fluctuations can be neglected; secondly, the signal must be an analytical function of time. Strictly speaking, the last condition is in contradiction with the physical statement of the measurement problem. An analytical function can be reconstructed over the entire time from the segment data which is known as a result of direct observations. This pertains to any time in the past as well as in the future. Therefore, real signals cannot be represented by analytical functions. However, any continuous function can be approximated over a finite interval by an analytical function with a bounded spectrum; this makes it possible to extrapolate (to a certain degree of approximation) the initial function outside the interval for which the function is known as a result of direct observations. Since a more detailed discussion of this topic is beyond the scope of the book, we can only note here that for an approximating function with a sufficiently wide frequency band, the extrapolation renders a satisfactory approximation in the time interval which substantially exceeds the initial measurement interval (see Slepian & Pollack [41], pt. 1, and also Khurgin & Yakovlev [71], p. 137).

The increase in frequency resolution is essentially related to such an extrapolation. The engineering realization of this resolution is exceedingly difficult, but in principle the only limitation is the measurement accuracy of the output signal, *i.e.*, in the final analysis, the accuracy is determined by the level of the noise fluctuations.

A statistical treatment of the resolution problem is particularly applicable to frequency measurement and will be discussed in the next chapter. Moreover, it will be shown that the resolution not only depends on the width of the impulse response of the measuring device (as in the classical case) but also on the signal-to-noise ratio. The above consideration pertains to the desired situation of a high signal-to-noise ratio. It is evident that by carrying this to the ultimate, there is no limitation to the resolution. It is also clear from the above that the widely accepted viewpoint of resolution being inversely proportional to the analysis time is based on the Rayleigh definition of resolution. This definition considers only one aspect (it is true that this is the most important aspect) of the problem and, strictly speaking, other factors which are related to the accuracy of measurement, signal energy, etc., must also be considered. This does not deny that in many important problems, the Rayleigh criterion proves to be adequate, and that its use is fully justified.

[§16] MEASUREMENT OF INSTANTANEOUS FREQUENCY

In measuring the range rate of a target, not only the spectral frequencies but also the instantaneous frequency may be used. The *instantaneous frequency* is a localized characteristic; it is defined for every instant of time and can change from one instant to another; therefore, for measuring purposes, long-duration time intervals are not required.* However, some measuring

* The instantaneous frequency is understood to be the phase derivative of the sophisticated signal which, in turn, is determined from the Hilbert transform of a real oscillation (see Gonorovskii [13], p. 664, Middleton [48], Vol. 1, p. 102, and also §18). For a narrow-band, oscillating signal such a definition coincides with that generally accepted.

methods of instantaneous frequency have a limitation similar to the uncertainty relationship.

As an example, let us consider a method in which a resonant device (wavemeter) is used to measure the frequency components of an FM signal. By tuning the wavemeter to any frequency ω and supplying an FM signal to the input, a signal spike will appear at the output when the instantaneous frequency runs through ω_0. By retuning the wavemeter and recording the time when the spike appears, it is possible to determine the frequency function of the signal under investigation.

Each measurement consists essentially of two operations: first, the wavemeter is tuned to the chosen frequency (strictly speaking, the wavemeter is adjusted to detect only the chosen frequency), and then the time when the output spike appears is measured. It is not difficult to notice that the requirements for high accuracy of such a measurement are contradictory. In order to increase the frequency resolution, it is necessary to narrow the wavemeter's bandwidth, *i.e.*, to increase the circuit Q. However, the narrower the bandwidth, the longer the duration of the "ringing" in the wavemeter, *i.e.*, the longer the duration of the current spike at the output. The time error depends on the duration of the "ringing" and, consequently, all other things being equal, this error increases as the frequency resolution is increased.

Let us discuss this dependence in greater detail. Disregarding the instrument errors (caused by the backlash in the micrometer, parallax, etc.), the tuning error of the wavemeter depends on the width of its resonance curve. By readjusting the wavemeter and recording the output level (with a constant frequency input) for each adjustment, a measurement is essentially made of the wavemeter amplitude-frequency characteristic. The wider the characteristic (thus, the more gradual its slope), the greater the calibration error.

Now let us consider the time measurement. The duration of the wavemeter's output signal is a function of two factors: the time required for the instantaneous frequency to run through the pass band, and the time of stabilizing the oscillations (which depends on the wavemeter's characteristics). The duration of the output signal is longer than in the case of the frequency changing at an infinite rate, *i.e.*, where one of the factors determining the signal duration is equal to zero. To help clarify the ultimate potentialities of this measurement, we shall consider one particular case. It can be shown (*e.g.*, see [9], p. 152) that for a sufficiently high modulation rate,* a wavemeter's response to an FM signal is the same as its impulse

* More accurately, if the condition $d\omega_c(t)/dt \gg \Delta^2$ is satisfied, where Δ is the wavemeter passband and $\omega_c(t)$ is the instantaneous frequency.

response:

$$u_{\text{out}}(t) = h(t - t_0),$$

where t_0 is the time of the measurement.

The impulse response is the wavemeter's response to a δ-function and is related to the wavemeter's frequency characteristic by the Fourier transform

$$h(t) = \frac{1}{2\pi} \int_{-\infty}^{\infty} H(\omega)\, e^{j\omega t}\, d\omega .$$

Thus, in calibrating the wavemeter, the frequency characteristic, $H(\omega)$, serves as the equipment function, and in measuring time the impulse response, $h(t)$, is related to $H(\omega)$ by means of the Fourier transform. Since these functions are Fourier transform pairs, the following is the uncertainty relationship [similar to (12–7)] for measuring the instantaneous frequency:

$$\sigma_\omega \sigma_t = \mu \geqslant \tfrac{1}{2}. \tag{16–1}$$

If σ_ω and σ_t are the wavemeter's frequency and time resolution, this result follows rigorously from the conclusions of §12. However, from the Rayleigh criterion, the values of these parameters are directly related to the measurement errors and then (16–1) characterizes the resolution (selectivity) of the measurement. It is clear from the above discussion that such an assumption is arbitrary; the resolutions are essentially the standard deviations of the measurements.

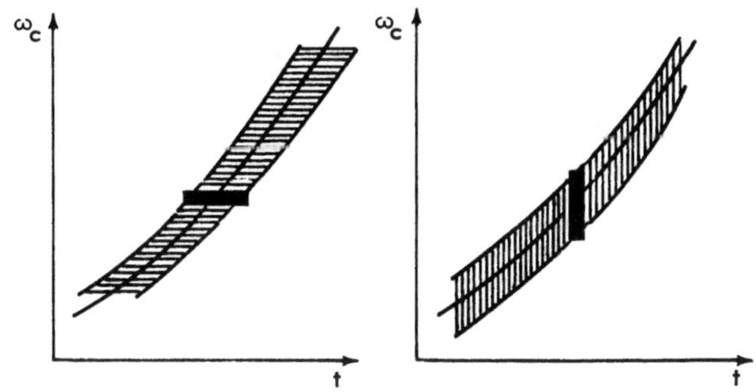

Fig. 24

Figure 24 shows two cases illustrating the uncertainty relationships for the measurement of instantaneous frequency of an FM signal. In the first case the time measurement resolution is large, and the resolution of the frequency measurement is small. Conversely, in the second case the resolution of the time measurement is small, while the resolution of the frequency measurement is large. The dark rectangle in each plot shows the uncertainty of the measurement, and the cross-hatched band is the probable error of the FM law measurement. It can only be said that the curve representing the modulation law is contained within this band. It is clear from Figure 24 that even substantial changes in the frequency and time resolutions (provided their product remains unchanged) do not result in a noticeable change of the total error. In order to decrease this error, it is necessary to decrease the *product* of the errors, *i.e.*, the quantity μ in (16–1). Such a possibility does follow from the general theory of measuring devices. It was shown previously that for a signal with sufficiently high energy, there exists, in principle, no limitations on the measurement resolution, and consequently, the standard deviations can be made very small. This fact is justified more rigorously in the next chapter on the basis of statistical interpretation. However, in reference to the measurement of instantaneous frequency, a sufficiently low value of μ in (16–1) can be realized even with present-day measurement technology.*

It was shown in §6 that to successfully compress an FM signal pulse, the modulation must be extremely accurate. The admissible standard deviation of the phase must satisfy the following condition:

$$\sigma_\psi \leqslant \lambda\pi \,. \tag{16–2}$$

For this reason it is necessary to control the change of frequency with an accuracy greater than this standard deviation.** However, it can be shown (see Appendix 2) that the standard deviation of the phase measurement is related to the frequency and time resolutions by

$$\sigma_\psi^2 \approx \sigma_\omega^2 T^2 + \sigma_t^2 \Omega^2 \,, \tag{16–3}$$

where T is the time duration of the FM signal and Ω its frequency deviation.

This relationship is very general and is applicable to any measurement method. Furthermore, considering (16–1), we find

$$\sigma_\psi^2 \approx \frac{T^2\mu^2}{\sigma_t^2} + \Omega^2\sigma_t^2 \,.$$

* Of course, what is meant here is that σ_ω and σ_t are the standard deviations of the measurements and not the durations of the wavemeter's resolution functions. In the latter case the condition $\mu \geqslant 1/2$ is absolutely correct.

** Even if the signal is formed with the "key-locked" system (see §7), it is still necessary to monitor the FM law since substantial distortion is introduced by power amplifiers. The monitoring of the instantaneous frequency is accomplished by a discriminator that compares the transmitter output signal with a "standard" formed from a linear filter. The error signal from the discriminator is used in a tracking system for tuning the frequency or phase (see, for example [40]).

If the value of μ is fixed, an increase in σ_t increases one term and decreases the other. There is an optimal value of σ_t which results in a minimum σ_ψ. After determining this minimum, we find

$$\sigma_\psi^2 \geqslant 2\mu T\Omega = 4\pi\mu m, \tag{16–4}$$

where $m = T\Omega/2\pi$ is the compression ratio.

Thus, the lower boundary of the standard deviation of the phase measurement depends only on μ and the signal parameter (compression ratio). Considering the allowable phase error from (16–2), it can be readily shown that for a given compression ratio, the value of μ is constrained by

$$\mu \leqslant \frac{\pi}{4} \frac{\lambda^2}{m}. \tag{16–5}$$

The variable λ^2 is the order of unity; in practice, FM signals are compressed by a factor of several hundreds and even several thousands. Consequently, the values of μ realized in practice are in the range 10^{-2} to 10^{-3}, and this is substantially lower than the "ultimate" of 1/2, which corresponds to the uncertainty principle. Thus, the experience of compressing FM signal pulses is a convincing confirmation of the fact that for high energy signals there is no limitation to the product of frequency and time accuracies.

[§17] THE GENERALIZED AMBIGUITY FUNCTION

Thus far our interest has been in the measurement and the measurement resolution of a target's range and range rate; however, we have not considered angular coordinate measurements. Also it must be understood, generally speaking, that the resolution of a radar is the capability to discriminate targets with respect to all parameters being measured. Obviously, for most radars, two targets having nearly the same range cannot be taken as one target if they are at substantially different azimuth angles. The resolution of the angular measurement increases with the directivity of the transmitting and receiving antenna patterns. Usually the spatial directivity of a radar is a function of the antenna characteristics, and the form of the signal does not have any noticeable effect on the angular measurement. This is why the study of angular measurements is independent from that of the range and range rate measurements; the former is a function of the antenna pattern, while the latter is a function of the signal structure. However, such an approach is not rigorous, since the antenna directional properties depend on the signal frequency (and thus, the signal form). For example, frequency scanning antennas, and multi-element slot antennas illustrate the above dependence.

Assume that the transmitter generates a continuous wave signal of constant amplitude, the frequency of which varies in a sawtooth fashion. The antenna beam will scan over a given angular interval at the expense of frequency modulating the signal, and the signal reflected from a target is no longer continuous. The duration of the reflected signal is a function of the length of time the beam points at the target, *i.e.*, ideally, the time corresponding to the width of the antenna pattern for a monochromatic signal. However, the instantaneous frequency of this pulse varies according to its modulation, and it is possible to compress the reflected signal in the receiver. As a result, the processed signal will have short duration. By measuring the time of the pulse's appearance and its frequency, it is generally possible to determine both the target's azimuth and range. Evidently, the design of a system to measure angles combines the extraction of the range and range rate data in one process, and all measurements are dependent on the modulation law.

In this section the *generalized ambiguity function*, which characterizes the radar's resolution of range, range rate, and two solid angles, is considered. This presentation will basically follow the work of Urkowitz, *et al.*, [62].*

Let us first consider the simplest case, *i.e.*, resolution of only one parameter, namely, the range. Assume that the reflected signals arrive from two nearby targets whose separation corresponds to the time interval Δt. To resolve these targets, it is necessary to measure some differences of the reflected signals, *e.g.*, the difference between $s(t)$ and $s(t-\Delta t)$, assuming all other signal parameters are equal. The following criterion is a possibility. Let us consider the integral

$$\mathcal{S}(\Delta t) = \int_{-\infty}^{\infty} |s(t) - s(t-\Delta t)|^2 \, dt. \tag{17–1}$$

At least for small Δt, the value of the integral increases as Δt increases; by introducing some threshold level, a decision can be made whether one or two targets are present, depending on whether $\mathcal{S}(\Delta t)$ is greater or smaller than the threshold level. Such a criterion assumes that the "inherent" properties of the signal defined by its modulation law are known, since the value of $\mathcal{S}(\Delta t)$ depends on the function $s(t)$. By analyzing the dependence of $\mathcal{S}(\Delta t)$ on the form of $s(t)$, it is possible to choose a modulation characteristic resulting in the necessary resolution. This is the main advantage of the criterion described above. However, in using such an approach, the effect of noise and other interfering factors are completely neglected, so that, strictly speaking, an objective resolution criterion is not obtained. In fact,

* Analogous results were obtained by Mayo, *et al.*, [47]; a more general statistical interpretation of resolution is given by Shirman [74].

with an absence of noise, it is possible to choose a threshold level arbitrarily low and, consequently, to resolve targets as close together as desired for any signal $s(t)$. Thus, the natural approach taken above for estimating resolution is equivalent, more or less, to the classical Rayleigh definition and does not contradict the principles presented in the previous sections.

It is not difficult to make the transition from (17–1) to the ambiguity function, more accurately, the transition to the signal's autocorrelation function which defines the range resolution. We have

$$\begin{aligned}\mathscr{S}(\Delta t) &= \int_{-\infty}^{\infty} [s(t) - s(t-\Delta t)]\,[s^*(t) - s^*(t-\Delta t)]\,dt \\ &= \int_{-\infty}^{\infty} |s(t)|^2\,dt + \int_{-\infty}^{\infty} |s(t-\Delta t)|^2\,dt \\ &\quad - \int_{-\infty}^{\infty} [s(t)\,s^*(t-\Delta t) + s^*(t)\,s(t-\Delta t)]\,dt \\ &= 2E\left(1 - \operatorname{Re}\frac{1}{E}\int_{-\infty}^{\infty} s(t)\,s^*(t-\Delta t)\,dt\right),\end{aligned}$$

where E is the total energy of the reflected signal.

It follows from this criterion that for better range resolution, the real part of the autocorrelation function,

$$\operatorname{Re} R(t) = \operatorname{Re}\frac{1}{E}\int_{-\infty}^{\infty} s(t')\,s^*(t'-t)\,dt', \tag{17–2}$$

must be as small as possible for $t \neq 0$, *i.e.*, Re $R(t)$ must be a pulse of short duration. It has been assumed thus far that such properties are not characteristic of the real part of the function $R(t)$ but of its magnitude. However, it can be readily shown that there is no essential difference between these two assumptions. In fact, when the signal

$$s(t) = a(t)\,e^{j\omega_0 t},$$

where $a(t)$ is the complex envelope, has a sufficiently high carrier frequency, the autocorrelation function is

$$R(t) = \frac{1}{E}\,e^{j\omega_0 t}\int_{-\infty}^{\infty} a(t')\,a^*(t'-t)\,dt'.$$

Introducing the notation,

$$r(t) = |r(t)|\, e^{j \arg r} = \frac{1}{E} \int_{-\infty}^{\infty} a(t')\, a^*(t' - t)\, dt',$$

so that $|r(t)| = |R(t)|$, we obtain

$$R(t) = |r(t)|\, e^{j(\omega_0 t + \arg r)} = |R(t)|\, e^{j(\omega_0 t + \arg r)}$$

and

$$\operatorname{Re} R(t) = |R(t)| \cos(\omega_0 t + \arg r).$$

Since, in this case, t corresponds to the distance between targets, $\omega_0 t$ is a very large number and significant changes in the distance (with respect to the wavelength) causes $\omega_0 t$ to change in the order of π. The arg r changes relatively slowly and, therefore, the only way to decrease Re $R(t)$ in a not too small time interval is to decrease $|R(t)|$. Thus, proceeding from the criterion (17–1), we determined, from the above discussion, the necessity of having the amplitude component of the autocorrelation function in the form of a sharp peak.

Quantitatively, the range resolution is a function of the width of this peak, which is given by the following:

$$\Delta = \frac{\int_{-\infty}^{\infty} |R(t)|^2\, dt}{|R(0)|^2}. \tag{17–3}$$

Representing $R(t)$ in terms of the signal spectrum, we obtain

$$\Delta = \frac{\int_{-\infty}^{\infty} |S(\omega)|^4\, d\omega}{\frac{1}{2\pi}\left[\int_{-\infty}^{\infty} |S(\omega)|^2\, d\omega\right]^2} = \frac{1}{F_{\mathrm{eff}}},$$

where F_{eff} is the effective signal spectrum width defined by

$$F_{\mathrm{eff}} = \frac{\left[\int_{-\infty}^{\infty} |S(\omega)|^2\, d\omega\right]^2}{2\pi \int_{-\infty}^{\infty} |S(\omega)|^4\, d\omega}. \tag{17–4}$$

It should be noted that a similar quantity was used in determining the compression ratio (see §4).*

A similar approach is used for the resolution of other parameters. If the reflected signals $s_1(t)$ and $s_2(t)$ differ due to the Doppler shift (because of different target velocities), the following integral is pertinent:

$$\mathscr{S}(\Omega) = \int_{-\infty}^{\infty} |s_1(t) - s_2(t)|^2 \, dt$$

$$= \int_{-\infty}^{\infty} |s(t)|^2 \, |1 - \mathrm{e}^{-j\Omega t}|^2 \, dt$$

$$= 2E\left(1 - \mathrm{Re} \frac{1}{E} \int_{-\infty}^{\infty} |s(t)|^2 \, \mathrm{e}^{-j\Omega t} \, dt\right).$$

Thus, in measuring target range rates, the resolving properties of a signal are characterized by the function

$$\chi(\Omega) = \frac{1}{E} \int_{-\infty}^{\infty} |s(t)|^2 \, \mathrm{e}^{-j\Omega t} \, dt = \frac{1}{2\pi E} \int_{-\infty}^{\infty} S(\omega + \Omega) \, S^*(\omega) \, d\omega \, .$$

The width of this function is a quantitative measure of the resolution,

$$\Delta = \frac{\int_{-\infty}^{\infty} |\chi(\Omega)|^2 \, d\Omega}{|\chi(0)|^2} = \frac{2\pi}{T_{\mathrm{eff}}},$$

where T_{eff} is the effective signal duration.

$$T_{\mathrm{eff}} = \frac{\left[\int_{-\infty}^{\infty} |s(t)|^2 \, dt\right]^2}{\int_{-\infty}^{\infty} |s(t)|^4 \, dt} . \tag{17–5}$$

* This definition of effective spectrum width (and also the analogous definitions of effective signal duration and antenna length used in the following) corresponds to the general definition of Zakai with $v = 4$ [cf. (13–4)].

For targets having different ranges and range rates, it is necessary to analyze the difference between the two signals:

$$s_1 = s(t - t_1)\,\mathrm{e}^{-j\Omega_1 t}$$

and

$$s_2 = s(t - t_2)\,\mathrm{e}^{-j\Omega_2 t}.$$

Denoting the differences of these signal parameters by $\Delta t = t_2 - t_1$ and $\Omega = \Omega_2 - \Omega_1$, we obtain an expression which characterizes the two-dimensional "spread" between the targets:

$$\mathscr{S}(\Delta t, \Omega) = \int_{-\infty}^{\infty} |s(t - t_1)\,\mathrm{e}^{-j\Omega_1 t} - s(t - t_1 - \Delta t)\,\mathrm{e}^{-j(\Omega_1 + \Omega)t}|^2\,dt.$$

From this expression, it is not difficult to obtain the Woodward ambiguity function,

$$\chi(t, \Omega) = \frac{1}{E} \int_{-\infty}^{\infty} s^*(t')\,s(t' - t)\,\mathrm{e}^{j\Omega t'}\,dt',$$

and to show that the resolution,

$$\Delta = \frac{\iint\limits_{-\infty}^{\infty} |\chi(t, \Omega)|^2\,dt\,d\Omega}{\chi^2(0, 0)},$$

does not depend on the signal form and equals 2π. The mathematical operations for this result have been carried out in §10. The resolution Δ is nothing more than the area of the effective cross section of the function $|\chi|^2$.

Angular measurements will now be discussed. Let us consider a receiving antenna having a linear aperture (see Fig. 25). If the target is sufficiently remote so that the wave received from the target in the θ direction is plane,

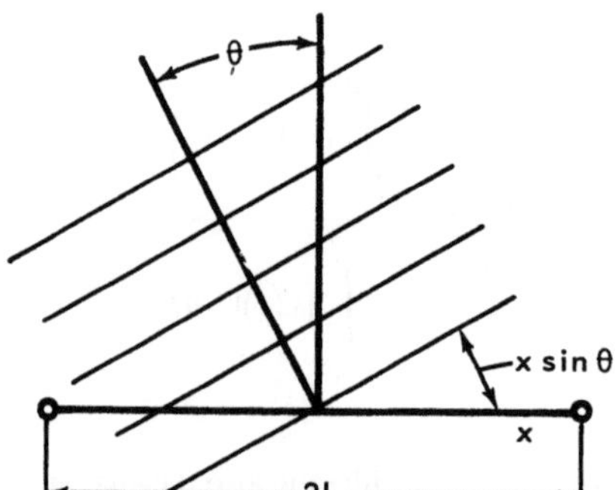

Fig. 25

the field along the aperture can be written as

$$s(x, t) = s\left(t - \frac{x}{c}\sin\theta\right),$$

where c is the velocity of light. The shifts in the signal's time of arrival, $(x/c)\sin\theta$, are due to the oblique incidence (see Fig. 25) of the wavefront. The currents excited along the aperture are collected in a horn or another type feed from which the output signal of the antenna is recorded. Moreover, the contributions from various increments of the antenna aperture differ due to the *aperture illumination function*, $I(x)$. Therefore, the signal received by an increment of antenna aperture from a target at angle θ has the form

$$s_x(t, \theta) = I(x)\, s\left(t - \frac{x}{c}\sin\theta\right). \tag{17–6}$$

The illumination function differs from zero only in the interval $-L \leqslant x \leqslant L$, *i.e.*, only over the physical dimension of the aperture. In general, $I(x)$ is a complex function having both amplitude and phase components. For convenience, let us introduce the generalized angular coordinate

$$u = \frac{\sin\theta}{c}$$

to compute the integrated difference of signals reflected from two targets located at the angles θ_1 and θ_2 (using u_1 and u_2):

$$\mathcal{S}(\Delta u) = \iint\limits_{-\infty}^{\infty} |s_x(t, u_1) - s_x(t, u_1 + \Delta u)|^2\, dx\, dt\,,$$

where $\Delta u = u_2 - u_1$.

It follows from (17.6) that

$$\begin{aligned}
\mathcal{S}(\Delta u) &= \iint\limits_{-\infty}^{\infty} |I(x)|^2\, |s(t - xu_1) - s(t - xu_1 - x\Delta u)|^2\, dx\, dt \\
&= \iint\limits_{-\infty}^{\infty} |I(x)|^2\, |s(t - xu_1)|^2\, dx\, dt \\
&\quad + \iint\limits_{-\infty}^{\infty} |I(x)|^2\, |s(t - xu_2)|^2\, dx\, dt \\
&\quad - 2\,\mathrm{Re} \iint\limits_{-\infty}^{\infty} |I(x)|^2\, s(t - xu_1)\, s^*(t - xu_1 - x\Delta u)\, dx\, dt\,.
\end{aligned}$$

It is readily seen that the first two integrals are the energies of the reflected signals appearing at the antenna output, and that the last integral defines the ambiguity function in terms of the angular coordinate:

$$\chi(u) = \iint\limits_{-\infty}^{\infty} |I(x)|^2 \, s(t) \, s^*(t - xu) \, dx \, dt . \tag{17–7}$$

Let us express this important function in a different manner. The integral with respect to t in (17–7) is the signal's autocorrelation function

$$R(xu) = \int\limits_{-\infty}^{\infty} s(t) \, s^*(t - xu) \, dt .$$

Therefore,

$$\chi(u) = \int\limits_{-\infty}^{\infty} |I(x)|^2 \, R(xu) \, dx .$$

Expressing the autocorrelation function in terms of the spectrum, $S(\omega)$, we obtain

$$\chi(u) = \frac{1}{2\pi} \iint\limits_{-\infty}^{\infty} |I(x)|^2 \, |S(\omega)|^2 \, e^{j\omega ux} \, d\omega \, dx .$$

The antenna pattern at any frequency ω can be represented by the form

$$G_\omega(u) = \int\limits_{-\infty}^{\infty} I(x) \, e^{j\omega ux} \, dx ,$$

where

$$\omega u = \frac{2\pi}{\lambda} \sin\theta .$$

Hence, it follows that

$$\int\limits_{-\infty}^{\infty} |I(x)|^2 \, e^{j\omega ux} \, dx = 2\pi \int\limits_{-\infty}^{\infty} G_\omega^*(u') \, G_\omega(u' + u) \, du' .$$

Therefore,

$$\chi(u) = \iint\limits_{-\infty}^{\infty} |S(\omega)|^2 \, G_\omega^*(u') \, G_\omega(u' + u) \, du' \, d\omega . \tag{17–8}$$

The angular resolution is determined using the common assumption that the signal has a sufficiently narrow frequency band. As a rule, it can be assumed that the antenna pattern varies insignificantly within the signal spectrum, *i.e.*,

$$G_\omega(u) \approx G_{\omega 0}(u) = G(u).$$

Then, the double integral (17–8) separates into two single integrals, and

$$\chi(u) = 2\pi E \int_{-\infty}^{\infty} G^*(u)\, G(u' + u)\, du'$$

$$= E \int_{-\infty}^{\infty} |I(x)|^2\, e^{j\omega_0 ux}\, dx$$

(E is the pulse energy). Thus, the angular resolution is

$$\Delta = \frac{\int_{-\infty}^{\infty} |\chi(u)|^2\, du}{|\chi(0)|^2} = \frac{2\pi}{\omega_0 L_{\text{eff}}} = \frac{1}{c}\frac{\lambda_0}{L_{\text{eff}}}, \tag{17–9}$$

where

$$L_{\text{eff}} = \frac{\left[\int_{-\infty}^{\infty} |I(x)|^2\, dx\right]^2}{\int_{-\infty}^{\infty} |I(x)|^4\, dx}$$

is the effective aperture length; λ_0 = the mean wavelength.

For small θ, where $\Delta u \approx \Delta\theta/c$, we obtain

$$\Delta\theta \approx \frac{\lambda}{L_{\text{eff}}}.$$

This result has a basic significance. It is usually assumed that the angular resolution depends on the width of the antenna pattern. However, from the foregoing discussion, it follows that the angular resolution is a function of the *effective aperture length* and not the antenna pattern width. For antennas with symmetrical aperture illuminations there is no contradiction here because the pattern width is inversely proportional to the effective aperture length. However, the antenna pattern may be broadened substantially due to phase errors in the aperture illumination, $I(x)$.* By selecting the proper

* These errors are of the same type as those occurring in the frequency modulation law for a compressed pulse (see §6).

phase distribution, it is possible to obtain various antenna patterns, including very wide ones. As an example, for a random, statistically independent phase distribution of $I(x)$, the resulting antenna pattern is "noise-like" without an explicit main lobe. Moreover, the signal is radiated into space randomly, *i.e.*, on the average in a non-directional manner, and it might seem that any angular resolution is not possible. However, by rotating such an antenna, there appears at the receiver output a signal that repeats the complicated, "noise-like" antenna pattern (a single target is implied here). Furthermore, by using an optimal filter matched to such a signal (of course, the filter parameters depend on the scanning velocity), we obtain a pulse of short duration, the position of which in time is a function of the target's angular position. Such a possibility of "compressing" the antenna pattern by means of signal processing is a substantial contribution to the general theory of radar resolution as related to antenna design.

In principle, this theory is also of interest because of the following somewhat unexpected phenomena. It is well known that under certain conditions it is possible to obtain the so-called superdirective antenna. This means that even for relatively small antennas, the pattern is very sharp, and thus the pattern width is substantially smaller than λ/L_{eff}. When first derived, this type of antenna was considered promising and much attention was given to its theory. Subsequently, it was established that superdirectivity is achieved at the expense of a rapidly oscillating aperture illumination, and the realization of such an antenna in practice is exceedingly difficult. In addition, superdirectivity is associated with storing high reactive energy around the antenna, and this precludes matching effectively the antenna to the receiver (at least over a sufficiently wide frequency band). As a result, superdirective antennas did not find wide use because of the engineering problems (*e.g.*, see Fel'd & Bakhrakh [64], Khurgin & Yakovlev [71], p. 160).

The above considerations show that for the processing methods discussed here, the angular resolution depends not on the antenna pattern width but on the antenna length. Therefore, superdirective antennas (even if they were realizable) do not have any substantial advantage over the ordinary antenna.*

Furthermore, it is not difficult to express the resolution of the various parameter combinations measured by a radar (angle and range; angle and range rate; two angles and range; etc.). All these cases are not discussed here but they are given in [62]. However, let us briefly consider the most general case where the target's azimuth, elevation, range, and range rate are measured. An examination of (17–6) shows that for this case it is necessary to

* A statistical analysis shows that superdirectivity may prove useful if noise sources are located only in discrete directions; this special case is considered by Shirman [74].

analyze the integrated difference between the two signals

$$s_1 = I(x, y)\, s(t - t_1 - xu_1 - yv_1)\, e^{j\Omega_1 t}$$

and

$$s_2 = I(x, y)\, s(t - t_2 - xu_2 - yv_2)\, e^{j\Omega_2 t},$$

where $I(x, y)$ is the two-dimensional field aperture illumination, $v = \frac{\sin\varphi}{c}$, and φ is the elevation angle.

Using the integral

$$\mathscr{S} = \iiint_{-\infty}^{\infty} |s_1 - s_2|^2\, dx\, dy\, dt,$$

we obtain the general ambiguity function which characterizes the resolution with respect to all coordinates:

$$\chi(t, \Omega, u, v) = \iiint_{-\infty}^{\infty} |I(x, y)|^2\, s(t')\, s^*(t' - t - ux - vy) \times e^{j\Omega t'}\, dt'\, dx\, dy = \frac{1}{2\pi} \iiint_{-\infty}^{\infty} |I(x, y)|^2\, S^*(\omega) \times S(\omega - \Omega)\, e^{j\omega(t + ux + vy)}\, d\omega\, dx\, dy. \quad (17\text{–}10)$$

In order to determine the resolution, which in this case is defined by

$$\Delta = \frac{\iiiint_{-\infty}^{\infty} |\chi(t, \Omega, u, v)|^2\, dt\, d\Omega\, du\, dv}{|\chi(0, 0, 0, 0)|^2}, \quad (17\text{–}11)$$

we must perform some mathematical operations. Writing the square of the magnitude in the form (this is done to carry out the integration over all variables)

$$|\chi|^2 = \chi\chi^* = \int |I(x_1, y_1)|^2\, |I(x_2, y_2)|^2\, S^*(\omega_1)\, S(\omega_2)\, S(\omega - \Omega)\, S^*(\omega_2 - \Omega) \times e^{j[(\omega_1 - \omega_2)t + (\omega_1 x_1 - \omega_2 x_2)u + (\omega_1 y_1 - \omega_2 y_2)v]}\, dx_1\, dx_2\, dy_1\, dy_2\, d\omega_1\, d\omega_2,$$

we first integrate $|\chi|^2$ with respect to t. Considering that

$$\int_{-\infty}^{\infty} e^{j(\omega_1 - \omega_2)t}\, dt = 2\pi\delta(\omega_1 - \omega_2),$$

and making use of the δ-function, we can integrate $|\chi|^2$ with respect to ω_1. As a result, we obtain

$$\int |\chi|^2 \, dt = \frac{1}{2\pi} \int |I(x_1, y_1)|^2 \, |I(x_2, y_2)|^2 \, |S(\omega)|^2 \, |S(\omega - \Omega)|^2 \times e^{j[(x - x_2)\omega u + (y_1 - y_2)\omega v]} \, dx_1 \, dx_2 \, dy_1 \, dy_2 \, d\omega .$$

Now using

$$\int_{-\infty}^{\infty} e^{j(x_1 - x_2)\omega u} \, du = \frac{2\pi}{\omega} \, \delta(x - x_2),$$

we can integrate with respect to u, then with respect to x_1, and similarly with respect to v and y. We now find

$$\int |\chi|^2 \, dt \, du \, dv = 2\pi \int |I(x, y)|^4 \, \frac{|S(\omega)|^2}{\omega^2} \, |S(\omega - \Omega)|^2 \, dx \, dy \, d\omega .$$

Finally, integrating with respect to Ω and noting that the quadruple integral reduces to two double integrals, we find from (17–11)

$$\Delta = \frac{8\pi^3}{A_{\text{eff}}} \, \frac{\displaystyle\iint_{-\infty}^{\infty} \frac{|S(\omega)|^2}{\omega^2} \, |S(\omega - \Omega)|^2 \, d\omega \, d\Omega}{\left[\displaystyle\int_{-\infty}^{\infty} |S(\omega)|^2 \, d\omega\right]^2}, \qquad (17\text{–}12)$$

where

$$A_{\text{eff}} = \frac{\left[\displaystyle\iint_{-\infty}^{\infty} |I(x, y)|^2 \, dx \, dy\right]^2}{\displaystyle\iint_{-\infty}^{\infty} |I(x, y)|^4 \, dx \, dy}$$

is the effective aperture of the antenna.

For narrow-band signals, (17–12) can be simplified. If $S(\omega)$ differs from zero only in a sufficiently narrow band around ω_0, then it can be assumed that $1/\omega^2 = 1/\omega_0^2$; then

$$\Delta = \frac{8\pi^3}{A_{\text{eff}} \omega_0^2} = \frac{2\pi}{c^2} \, \frac{\lambda_0^2}{A_{\text{eff}}} . \qquad (17\text{–}13)$$

This result can be interpreted as follows. It has been shown that the azimuth resolution equals $\lambda_0 / c L'_{\text{eff}}$, where L'_{eff} is the effective dimension of the antenna in the horizontal direction. Similarly, for the elevation angle, $\Delta = \lambda_0 / c L''_{\text{eff}}$, where L''_{eff} is the effective vertical dimension. In addition, the range and range rate resolutions do not depend on the signal form and equal 2π. Therefore, if $A_{\text{eff}}=L'_{\text{eff}}L''_{\text{eff}}$ [rigorously, this is only true for the special case when the field illumination has the form $T(x, y)=I_1(x)\, I_2(y)$], the total resolution with respect to the four coordinates equals the product of the four partial resolutions. This is a general rule for determining the "combined" resolution of a narrow-band signal. For the composite resolution of the coordinates being measured (azimuth and range; azimuth, elevation, and range; etc.), it is possible to calculate the resolution of each coordinate separately and then multiply them. However, it should be remembered that this rule is only suitable for the important, but nevertheless special, case of narrow band signals. If the signal frequency spread is so wide that the antenna pattern varies substantially for various components of the spectrum, the relationship is more complicated than (17–12). These special relationships for various combinations of measured coordinates are given in [62], where it is shown that in many cases there is an optimal band of frequencies for any minimum resolution.

3

The Ambiguity Function in the Statistical Theory of Radar

[§18] SAMPLING SPACE

The time function representing the reflected radar signal, which arrives at the receiver input mixed with noise, can be represented by points or vectors in a certain abstract space. This representation makes it possible to simplify the formulations of the probability relationships and, therefore, is exceedingly useful in the statistical theory.

Any arbitrary real signal function $s(t)$ which has finite energy,*

$$E = \int s^2(t)\, dt,$$

can be expanded into a generalized Fourier series in a complete system of orthogonal functions

$$s(t) = s_1 \cdot f_1(t) + s_2 \cdot f_2(t) + \cdots + s_m \cdot f_m(t). \qquad (18\text{–}1)$$

It is assumed that the expansion is limited to m terms, *i.e.*, all the s_k coefficients equal zero for $k > m$. Subsequently, it will be possible to eliminate this limitation by having $m \to \infty$. The coefficients s_k are arbitrary and the quantity m is the number of degrees of freedom of the function $s(t)$. As $m \to \infty$, the number of degrees of freedom increases without bound, and then the function $s(t)$ encompasses all physically realizable signals.**

Since the base functions, $f_k(t)$, are orthogonal, *i.e.*,

$$\int f_k(t) f_l(t)\, dt = \delta_{kl} = \begin{cases} 1 & \text{for} \quad k = l, \\ 0 & \text{for} \quad k \neq l, \end{cases} \qquad (18\text{–}2)$$

* If not otherwise specified, the integration limits are assumed to be infinite.

** It should be noted that this condition of $m \to \infty$ is not compulsory. It was shown by Landau and Pollack ([41], pt. III) that any function approximately bounded in both frequency and time has, to a certain degree of approximation, a finite number of degrees of freedom. In fact, these are the functions which can be used as representations of physical signals.

the coefficients of the expansion are given by

$$s_k = \int s(t) f_k(t)\, dt .$$

The set of coefficients $\mathbf{s} \equiv (s_1, s_2, \ldots, s_m)$ *which uniquely defines the function* $s(t)$ *is called the sampling, and the set of possible* $\mathbf{s}$ *is called the sampling space.* This space has m dimensions which is the same as the number of degrees of freedom of the signal. The signal $s(t)$ can also be described as m-dimensional. Imagine that the number s_k is plotted along its axis in the sampling space. The set of coefficients, $\mathbf{s}$, is then represented by a point in space or by an equivalent multi-dimensional radius-vector from the origin to the point. *Thus, every m-dimensional function has a corresponding vector in m-dimensional space.*

Parceval's theorem provides the following representation for the signal energy:

$$E = \int s^2(t)\, dt = \sum_{k=1}^{m} s_k^2 = |\mathbf{s}|^2 . \tag{18–3}$$

The energy E is represented by the square of the magnitude of the vector $\mathbf{s}$ and (18–3) is the generalized Pythagorean theorem.

Now let us consider the representation of noise in the sampling space. It is assumed that the noise is stationary and Gaussian, so that the probability density of a given noise level, n, at any arbitrary time t is

$$p(n) = \frac{1}{\sqrt{2\pi\sigma}}\, e^{-\frac{n^2}{2\sigma^2}},$$

where σ is the rms noise level (its effective intensity).

For random noise, $n(t)$, we have, from the general theory,

$$n(t) = \sum_{k=1}^{m} n_k f_k(t),$$

$$n_k = \int n(t) f_k(t)\, dt .$$

Considering the last integral as the limit of the summation and taking into account the additivity of random variables with normal distributions, it can be readily shown that the coefficients n_k also have normal distributions; moreover, the mean value of these coefficients $\overline{n_k} = 0$. The standard deviation σ_k^2 can be found in the following manner. We have

$$\sigma_k^2 = \overline{n_k^2} = \iint \overline{n(t)\, n(t')}\, f_k(t) f_k(t')\, dt\, dt' .$$

The mean value of the product $\overline{n(t)\,n(t')}$ is, by definition, the autocorrelation function of the noise,

$$\overline{n(t)\,n(t')} = R(t - t'),$$

Moreover, for white noise with a uniform power spectrum, N (see Helstrom [70], p. 50), we have

$$R(t - t') = N\delta(t - t').$$

Therefore, using (18–2), we find

$$\overline{n_k^2} = N \iint \delta(t - t') f_k(t) f_k(t')\, dt\, dt' = N \int f_k^2(t)\, dt = N.$$

Consequently, the random noise coefficients n_k have normal distributions which are identical for all k:

$$p(n_k) = \frac{1}{\sqrt{2\pi N}}\, e^{-\frac{n^2 k}{2N}}. \qquad (18\text{–}4)$$

An important property of noise is the *statistical independence* of the n_k coefficients. In order to prove this statement, it is necessary to calculate the covariance of the random variables n_k and n_l for $l \neq k$. We have

$$\begin{aligned} K_{kl} = \overline{n_k n_l} &= \iint \overline{n(t)\,n(t')} f_k(t) f_l(t')\, dt\, dt' \\ &= N \int \delta(t - t') f_k(t) f_l(t')\, dt\, dt' \\ &= N \int f_k(t) f_l(t)\, dt = 0, \end{aligned}$$

where the orthogonality between the functions $f_k(t)$ and $f_l(t)$ is taken into account.

The fact that the covariance equals zero denotes a *linear independence* of the random variables; in general, it cannot be asserted that these variables are statistically independent. However, when these variables have normal distributions, they are statistically independent, and thus the condition $K_{kl}=0$ means that n_k and n_l are independent of each other (see Davenport & Root [23], p. 149). Consequently, when white noise is expressed by any orthogonal system $\{f_k(t)\}$, the coefficients n_k are independent random variables having the normal distribution (18–4).*

* If the noise is Gaussian but not white and has an autocorrelation function $R(t)$, it is also possible to express the noise by a special $\{f_k(t)\}$ system which depends on $R(t)$. This is the *Karhunen-Loéve expansion* (see Davenport & Root [23], p. 96, and also Helstrom [70], p. 95).

Since the noise, $n(t)$, is fully defined by a set of random coefficients n_k, *i.e.*, by a random point in the sampling space, its representation can be interpreted as a multivariate random variable with the distribution

$$p(n(t)) = p(n_1, n_2, \ldots, n_m) = p(n_1)\, p(n_2) \ldots p(n_m)$$

$$= \frac{1}{(2\pi N)^{\frac{m}{2}}}\, e^{-\frac{\sum\limits_{k=1}^{m} n^2{}_k}{2N}}. \tag{18–5}$$

The transition from random functions of time into a multivariate random variable is, in fact, the principal result obtained by using the sampling space.

It is clear from (18–5) that for the geometrical representation, the noise fills a hypersphere with the center at the origin of the sampling space. If, for example, it is assumed that the random variable n_k does not exceed the $3\sigma_k$ value, then the radius of the hypersphere is $3\sqrt{N}$.

Let the received signal at the input of the receiver, $x(t)$, consist of a reflected signal, $s(t)$, and a noise, $n(t)$ (all functions are real):

$$x(t) = s(t) + n(t).$$

Expanding each component in the base system, the coefficients obtained are in the following form:

$$x_k = s_k + n_k, \qquad k = 1, 2, \ldots, m.$$

In the sampling space, the vector $\mathbf{x}$ is the sum of the constant vector $\mathbf{s}$ and a random vector $\mathbf{n}$. From (18–5), the distribution of the vector $\mathbf{x} \equiv (x_1, x_2, \ldots, x_n)$ has the form

$$p(x(t)) = p(\mathbf{x}) = \frac{1}{(2\pi N)^{\frac{m}{2}}}\, e^{-\frac{\sum\limits_{k=1}^{m} (x_k - s_k)^2}{2N}}. \tag{18–6}$$

The geometrical points representing this expression in the sampling space form a hypersphere, the center of which is at $\mathbf{s} = (s_1, s_2, \ldots, s_n)$.

Let us apply the complex notation to the signals used previously. For a complex signal, the real component, $s_1(t)$, is complemented by an imaginary component, $s_2(t)$:

$$s(t) = s_1(t) + js_2(t) = a(t)\, e^{j\varphi(t)},$$

where $s_2(t)$ may be, generally speaking, arbitrarily selected. However, if it is required that the magnitude of the signal function $|s(t)| = a(t)$, the signal envelope, and that the argument of the exponential $e^{j\varphi(t)}$ characterize the signal's phase structure, the choice of the imaginary component cannot be arbitrary: $s_2(t)$ is uniquely defined by $s_1(t)$ by means of the Hilbert transform (*e.g.*, see Gonorovskii [13], p. 664, and Middleton [48], Vol. I, p. 103).

With $s_2(t)$ defined in such a manner, let us expand it in the base system. As a result, the coefficients s_k of the complex signal $s(t)$ also become complex:

$$s_k = s_{1k} + js_{2k}, \qquad k = 1, 2, \ldots, m,$$

and the sampling space can be considered as consisting of two subspaces, each of which has m-dimensions. By virtue of the unique relationship between the real and imaginary signal components, the representative points in the subspaces are related, so that the signal still has m degrees of freedom instead of $2m$.

The energy of a complex signal function

$$E = \int |s(t)|^2 \, dt = \sum_{k=1}^{m} |s_k|^2 = \sum_{k=1}^{m} (s_{1k}^2 + s_{2k}^2) = E_1 + E_2$$

is equally distributed between the real and imaginary components (such is the property of the Hilbert transform), and therefore,

$$E = 2E_1 . \tag{18–7}$$

Thus, *the energy of a complex signal is always equal to double the energy of the real component*. Consequently, by using complex signal functions, (18–5) and (18–6) take the form

$$p(n(t)) = p(\mathbf{n}) = \frac{1}{(2\pi N)^{\frac{m}{2}}} e^{-\frac{\sum_{k=1}^{m} |n_k|^2}{4N}}, \tag{18–5'}$$

$$p(x(t)) = p(\mathbf{x}) = \frac{1}{(2\pi N)^{\frac{m}{2}}} e^{-\frac{\sum_{k=1}^{m} |x_k - s_k|^2}{4N}}. \tag{18–6'}$$

Equation (18–3), expressing the signal energy in terms of the vector components of the sampling space, is also valid here. It should be noted that, as before, N is the noise density for positive frequencies and accounts for only the real part of the noise. The addition of the imaginary part would result in doubling the density. However, if the noise energy is assumed to be distributed over all frequencies (both positive and negative), N can be interpreted as the noise density of the complex signal.

In the following, we shall encounter random variables which are linear transforms of the noise:

$$z = \int g(t)\, n(t)\, dt. \tag{18–8}$$

Here $g(t)$ and $n(t)$ are complex functions formed from the real functions by means of the Hilbert transform.

Expanding each of these functions in the basic system and considering (18–2), we obtain

$$z = \sum_{k=1}^{m} g_k n_k = g\mathbf{n}. \tag{18–8'}$$

In the geometrical analogy, the scalar product of the vectors corresponds to the linear operator (18–8); this fact is reflected by the above notation. Since $g_k = g_{1k} + jg_{2k}$ and $n_k = n_{1k} + jn_{2k}$, we have

$$z = x + jy = \left(\sum_{k=1}^{m} g_{1k}n_{1k} - \sum_{k=1}^{m} g_{2k}n_{2k}\right) + j\left(\sum_{k=1}^{m} g_{2k}n_{1k} + \sum_{k=1}^{m} g_{1k}n_{2k}\right).$$

The coefficients n_{1k} and n_{2k} are random with normal distributions. Therefore, in the last equation, each of the sums is a random variable with a normal distribution. It can be shown by means of the Hilbert transform that these sums are not correlated and, consequently, are independent. Thus, the real and imaginary components of z (the quadrature components),

$$x = \operatorname{Re} z \quad \text{and} \quad y = \operatorname{Im} z$$

have a normal distribution. The corresponding standard deviations are determined as follows:

$$\sigma_x^2 = \overline{x^2} = \sum_{k=1}^{m} \left(g_{1k}^2 \overline{n_{1k}^2} + g_{2k}^2 \overline{n_{2k}^2}\right) = N \sum_{k=1}^{m} \left(g_{1k}^2 + g_{2k}^2\right) = N \sum_{k=1}^{m} |g_k|^2 = N\,|\mathbf{g}|^2 .$$

Similarly,

$$\sigma_y^2 = \overline{y^2} = N\,|\mathbf{g}|^2 .$$

Also,

$$\overline{|z|^2} = \overline{x^2} + \overline{y^2} = 2N\,|\mathbf{g}|^2 . \tag{18–9}$$

It is noted that from (18–8') that

$$\overline{|z|^2} = \overline{|\mathbf{gn}|^2} = |\mathbf{g}|^2\, \overline{|\mathbf{n}|^2} = |\mathbf{g}|^2 \left(\overline{\mathbf{n}_1^2} + \overline{\mathbf{n}_2^2}\right) = 2N\,|\mathbf{g}|^2 .$$

Since they can always be rigorously justified, similar transformations will also be used in the subsequent material.

These results can be readily interpreted by using the geometrical analogy. The scalar product of the vectors **n** and **g** is proportional to the product of the magnitude of the projection of **n** on **g** times the magnitude of **g**. Since the noise uniformly fills a hypersphere, the center of which is at the origin, the noise projections in all directions are identical; more precisely, the statistical properties of noise are invariant with respect to the projection direction. This situation exists for (18–9): the mean of z depends on the magnitude and not on the direction of the vector **g**.

We shall illustrate another point. Let $g(t)$ also be a function of another parameter so $g(t)=g(t,\alpha)$. From (18–8),

$$z(\alpha) = \int g(t,\alpha)\, n(t)\, dt\,.$$

For every type of realizable noise $n(t)$, the function $z(\alpha)$ has the usual meaning. However, if the type of noise is a random variable, then $z(\alpha)$ is also a random variable and the *functional dependence* on α vanishes. This means that for any given α, the function $z(\alpha)$ is arbitrary. The only remaining relationship is the *statistical dependence* of the function z on the parameter α, *i.e.*, the probability distribution, $p(z)$, has α as a parameter. For the random variable case, the notation $z(\alpha)$ is only a convention; it means that the statistical characteristics of z depend on α.

[§19] PRINCIPLE OF MAXIMUM LIKELIHOOD

Some basic problems encountered in radio-astronomical observations are discussed below.

The reflected signal at the receiver input is mixed with noise. After analyzing the received signal, the observer must answer the following questions:

(1) Is the received signal due to noise mixed with a reflected signal, or is it noise alone? (The detection problem.)

(2) If there is reflected signal present, what is the number of targets (one or several and, specifically, how many) beingobserved? (The resolution problem.)

(3) What target parameters are being measured, *i.e.*, range, range rate, etc.? (The parameter measurement problem.)

The presence of random fluctuations (noise) introduces an uncertainty. By analyzing the signals, it is only possible to determine the likelihood of one or another hypothesis based on the presence, quantity, and type of target parameters; evidently, the observer must base his decision on a hypothesis related to minimum risk. One must decide on some criterion to ensure sufficient reliability for a large number of observations. The majority of the known criteria can be reduced to the maximum likelihood criterion considered below.

The reflected signal $s(t)$ depends on the parameters of the target, *i.e.*, its range, range rate, surface characteristics, etc. Let us denote these parameters by $\alpha_1, \alpha_2, \ldots, \alpha_\mu$, and represent them by a multi-dimensional vector $\boldsymbol{\alpha} \equiv (\alpha_1, \alpha_2, \ldots, \alpha_\mu)$. In order to emphasize the relationship between the above parameters and the reflected signal, the signal is expressed by the form

$$s = s(t; \boldsymbol{\alpha}) \equiv s(t; \alpha_1, \alpha_2, \ldots, \alpha_\mu). \tag{19–1}$$

One of the parameters, for example α_1, may denote the number of targets. This parameter takes on only discrete values ($\alpha_1 = 1, 2, 3, \ldots$); other parameters can be either continuous or also discrete.

The detection, resolution, and measurement task is reduced, in the general case, to a multi-alternate decision, in which the observer is asked to establish the presence of specific targets that are, in turn, characterized by one or another set of parameters. The classes can be either discrete or, in the limiting case, continuous where even a small change in the parameter transfers a target into another class. Considering the continuous case, we identify the set of target classes with a set of vectors $\boldsymbol{\alpha}$. Consequently, to classify the detected signal requires measuring the parameters and resolving them (besides separating the various parameters, it is also possible to separate the individual targets).

Thus, let $s(t; \boldsymbol{\alpha})$ be a reflected signal expressed in terms of the target parameters $\boldsymbol{\alpha} = (\alpha_1, \alpha_2, \ldots, \alpha_\mu)$. The received signal at the input is distorted by noise and differs from the reflected signal $s(t; \boldsymbol{\alpha})$:

$$x(t) = s(t; \boldsymbol{\alpha}) + n(t).$$

Even for known target parameters, it is not possible to predict accurately the input signal; it is only possible to determine the *conditional probability* that the received signal will have the form $x(t)$ for a given set of target parameters. This conditional probability is denoted by $p(\mathbf{x}|\boldsymbol{\alpha})$. In accordance with the previous discussion, the signal $x(t)$ is represented by the multi-dimensional vector $\mathbf{x} = (x_1, x_2, \ldots, x_m)$ in the sampling space.

Conversely, if the signal $x(t)$ is received at the input, the target parameters also cannot be determined explicitly from the signal form. The target parameters can only be estimated to some degree of approximation; this estimate is denoted by the vector $\hat{\boldsymbol{\alpha}} \equiv (\hat{\alpha}_1, \hat{\alpha}_2, \ldots, \hat{\alpha}_\mu)$. The vector of these estimates, $\hat{\boldsymbol{\alpha}}$, is a multi-dimensional random variable which depends on the random noise contained in the signal $x(t)$. Finding a rule by which the estimate of the target parameters, $\mathbf{x}$, is obtained for every input signal $x(t)$, is the technique of estimating the target parameters.

The principle of the maximum likelihood is based on the assumption that the received signal $\mathbf{x}$ most likely corresponds to the target parameters $\boldsymbol{\alpha}$ for which the conditional probability $p(\mathbf{x}|\boldsymbol{\alpha})$ is maximum. *The value of* $\boldsymbol{\alpha}$ *for which the conditional probability* $p(\mathbf{x}|\boldsymbol{\alpha})$ *is maximum is chosen as the estimate* $\mathbf{x}$ *for the received signal.*

It should be noted that here the probability $p(\mathbf{x}|\boldsymbol{\alpha})$ is not a function of the multi-dimensional random variable vector $\mathbf{x}$, but is rather a function of the parameter vector $\boldsymbol{\alpha}$ (which is not necessarily random) with a fixed value of $\mathbf{x}$. In order to emphasize this distinction, a special notation is introduced:

$$p(\mathbf{x}|\boldsymbol{\alpha}) = \lambda_x(\boldsymbol{\alpha}),$$

and the function $\lambda_x(\boldsymbol{\alpha})$ is called the *likelihood function.* To make the estimate requires finding the maximum of the function:

$$\max_{\boldsymbol{\alpha}} \lambda_x(\boldsymbol{\alpha}) = \lambda_x(\hat{\boldsymbol{\alpha}}). \tag{19–2}$$

For continuous parameters, the problem reduces to finding the solution to the system of equations,

$$\frac{\partial \lambda_x}{\partial \alpha_i} = 0, \quad i = 1, 2, \ldots, \mu;$$

moreover, the selected solution corresponds to the largest maximum of λ_x (*i.e.*, the highest peak is chosen). For discrete parameters, it is necessary to calculate $\lambda_x(\boldsymbol{\alpha})$ for the admissible discrete values of $\boldsymbol{\alpha}$ and to select the largest λ_x.

Although this technique leads to the likelihood of the estimate vector $\hat{\boldsymbol{\alpha}}$, it is frequently not sufficient for the following reason. The received signal $x(t)$ may contain no reflected signal and be due to noise alone. Thus, in making the decision, the observer must compare $p(\mathbf{x}|\hat{\boldsymbol{\alpha}})$, the probability that the received signal contains a target reflection having parameters $\hat{\boldsymbol{\alpha}}$, with $p_0(\mathbf{x})$, the probability that the same signal is due to noise alone. For this purpose, the *likelihood ratio,*

$$\Lambda_x(\hat{\boldsymbol{\alpha}}) = \frac{p(\mathbf{x}|\hat{\boldsymbol{\alpha}})}{p_0(\mathbf{x})} = \frac{\lambda_x(\hat{\boldsymbol{\alpha}})}{p_0(\mathbf{x})}, \tag{19–3}$$

must be calculated, and the decision is made that a target with parameters $\hat{\boldsymbol{\alpha}}$ is present when this ratio exceeds a certain threshold. Otherwise, the decision is that no target is present. Since the vector $\hat{\boldsymbol{\alpha}}$, which satisfies condition (19–2), maximizes the likelihood ratio, the multi-alternate detection criterion can be formulated in its final form as follows: *the decision that a target with* $\hat{\boldsymbol{\alpha}}$ *parameters is present is made if the maximum likelihood ratio,* $\Lambda_x(\hat{\boldsymbol{\alpha}})$, *ex-*

ceeds the chosen threshold:

$$\max_{\boldsymbol{\alpha}} \Lambda_x(\boldsymbol{\alpha}) = \Lambda_x(\hat{\boldsymbol{\alpha}}) \geqslant l(\hat{\boldsymbol{\alpha}}). \tag{19–4}$$

The dependence of the threshold on $\boldsymbol{\alpha}$ is based on many additional factors, as, for example, the *a priori* information on the probability of finding various types of targets and also the "cost" of making one or another kind of errors. Thus, under certain conditions, a failure to detect a high-speed target may prove to be more dangerous than a false alarm and, in turn, the latter may be more dangerous than a failure to detect a slow-flying target. These and similar factors can be approximately considered by adjusting the threshold $l(\boldsymbol{\alpha})$. At least for some comparatively simple cases, the dependence of the threshold on *a priori* probabilities and cost functions has been given detailed consideration in the literature (*e.g.*, see Shirman & Golikov [75], p. 15; Bakut, *et al.* [1], Vol. I, pp. 142, 158).

It is noted that the parameters estimated from the maximum likelihood criterion are approximations. This method is asymptotic in the sense that it is in good agreement with more general criteria based on the signal-to-noise ratio (which is frequently equivalent to the case of a low probability of a false alarm). One of the conditions for the applicability of criterion (19–4) is that the dependence of the threshold on $\boldsymbol{\alpha}$ be weak.

Usually, among the α_i parameters there are items which do not contain any target information (*e.g.*, the initial phase of the signal). The likelihood ratio must be computed from the preliminary averages of the parameters (not subject to measurement), and then the likelihood ratio is compared to the corresponding threshold (*e.g.*, see Shirman & Golikov [75], p. 73). However, by using the method of maximum likelihood, it is possible (without performing any averaging) to determine the optimum with respect to all parameters including the "useless" ones, and to disregard these "useless" parameters in classifying the target. Such a technique is usually simpler when the decision is made automatically by the equipment; for the not too rigorous conditions mentioned above, good results are obtained (Helstrom, [70], p. 197; Bakut, *et al.* [1], Vol. I, p. 49).

The relationships from the previous section can be used to obtain explicit expressions for the likelihood function and the likelihood ratio. Since the reflected signal depends on the target parameter vector $\boldsymbol{\alpha}$, the components of $\mathbf{s}$, the vector representing the reflected signal, are also a function of $\boldsymbol{\alpha}$. Therefore, from (18–6′), the likelihood function has the form

$$\lambda_x(\boldsymbol{\alpha}) = p(\mathbf{x}|\boldsymbol{\alpha}) = \frac{1}{(2\pi N)^{\frac{m}{2}}} e^{-\frac{L_x(\boldsymbol{\alpha})}{4N}}, \tag{19–5}$$

where

$$L_x(\boldsymbol{\alpha}) = \sum_{k=1}^{m} |x_k - s_k(\boldsymbol{\alpha})|^2 . \tag{19–6}$$

Similarly, the probability that random noise produces the signal $x(t)$ is given by (18–5′):

$$p_0(\mathbf{x}) = \frac{1}{(2\pi N)^{\frac{m}{2}}} \mathrm{e}^{-\frac{\sum\limits_{k=1}^{m} |x_k|^2}{4N}} .$$

Therefore, the likelihood ratio is

$$\Lambda_x(\boldsymbol{\alpha}) = \mathrm{e}^{-\frac{L_x(\boldsymbol{\alpha}) - \sum\limits_{k=1}^{m} |x_k|^2}{4N}} = \mathrm{e}^{-\frac{L_x(\boldsymbol{\alpha}) - E_x}{4N}} . \tag{19–7}$$

Using the Parceval theorem, (19–6) can be written in terms of the time variable. Thus,

$$L_x(\boldsymbol{\alpha}) = \int |x(t) - s(t;\boldsymbol{\alpha})|^2 \, dt = E_x + E_s - 2 \operatorname{Re} \int x(t) \, s^*(t;\boldsymbol{\alpha}) \, dt , \tag{19–8}$$

where E_x and E_s are the energies of signals $x(t)$ and $s(t)$, respectively. The signal $x(t)$ and the energy E_x are functionally independent of $\boldsymbol{\alpha}$ (there is only statistical dependence). Generally, the energy E_s is a function of the signal parameters, but for frequency and time shift variations, the energy E_s can be assumed to be constant. Thus, (19–8) is significantly dependent on the parameters of the reflected signal in only the last term; it can be readily seen that this term is maximum when the likelihood ratio is maximum.

The last term in (19–8) is only the *correlation function of the received signal $x(t)$ and the expected signal $s(t\!:\boldsymbol{\alpha})$*. This function is formed at the output of the filter matched to the expected signal; the principal portion of this function is the ambiguity function,

$$\chi(\boldsymbol{\alpha} - \boldsymbol{\alpha}_0) = \int s(t;\boldsymbol{\alpha}_0) \, s^*(t;\boldsymbol{\alpha}) \, dt , \tag{19–9}$$

where $\boldsymbol{\alpha}_0$ is the vector of the true target parameters. The latter function in (19–9) is obtained from (19–8) with noise absent.*

* For the reasons given in §17, the difference between the real part and the magnitude of the ambiguity function is frequently insignificant.

It can be readily concluded that the above method of reception, which is based on the use of matched filters, realizes the optimal processing procedure which follows from the maximum likelihood principle. Some refinements and concrete examples of this statement are given in subsequent sections of the book.

[§20] THE AMBIGUITY FUNCTION IN RELATION TO RESOLUTION

The principle of the maximum likelihood will now be applied to determining the resolution. It is assumed that two signals of known form, $Af(t)$ and $Bg(t)$, can be received (or not received) at the input. The complex functions $f(t)$ and $g(t)$ are known; moreover, $g(t)$ can be identical to $f(t)$ if shifted in time or frequency. It can be assumed without any loss in generality that these functions are normalized with respect to energy:

$$\int |f(t)|^2 \, dt = \int |g(t)|^2 \, dt = 1 . \tag{20–1}$$

The initial phases of the reflected signals are arbitrary and, therefore, the coefficients A and B are complex. Depending on the situation, the amplitude of each signal can take either of two values: it is zero in the absence of a target, and it is A_0 (or B_0) if a target is present. Thus, four hypotheses (alternatives) are possible:

hypothesis 1 —	$A=0,$	$B=0;$
hypothesis 2 —	$A=A_0 \, e^{j\theta_1},$	$B=0;$
hypothesis 3 —	$A=0,$	$B=B_0 \, e^{j\theta_2};$
hypothesis 4 —	$A=A_0 \, e^{j\theta_1},$	$B=B_0 \, e^{j\theta_2}.$

The observer must choose the most likely one.*

For this purpose, let us postulate that the received signal is the superposition of noise and the expected signals:

$$x(t) = n(t) + Af(t) + Bg(t) , \tag{20–2}$$

and that the most likely estimates of the amplitudes $\hat{A}_0$ and $\hat{B}_0$ are known. Furthermore, by comparing these estimates with the corresponding thresholds, the observer makes a decision on the presence of one or the other signal and, consequently, selects one of the hypotheses.

* This section basically follows Helstrom ([70], p. 267); however, signals with arbitrary phases are considered. More general cases are considered by Shirman [74], Root [52], and others.

First, we shall investigate the optimal criterion for making the estimates and then refine the choice of thresholds. In doing this, the role of the ambiguity function in the resolution phenomenon will be elucidated.

The most likely estimate results in a maximum of the ambiguity function; in accordance with (19–5), this requires a minimization of

$$L_x(A, B) = \int |x - Af - Bg|^2 \, dt .$$

Obviously, estimating the amplitudes A and B is equivalent to approximating the signal $x(t)$ by a linear combination of the functions $f(t)$ and $g(t)$ using the method of the least squares. By writing the above equation in the form

$$L_x(A, B) = \int (x - Af - Bg)(x^* - A^* f^* - B^* g^*) \, dt , \qquad (20\text{–}3)$$

it is noted that the complex conjugate quantities A and A^* (also B and B^*) can be considered here as independent parameters. This is associated with the fact that from the method of the maximum likelihood, the estimates are determined not only from the signal amplitudes but also from the phases which are not measured. Obviously, the determination of A and A^* is equivalent to finding the amplitude A_0 and the phase θ_1. Differentiation of (20–3) leads to a system of "linear" equations* with respect to the estimates $\hat{A}$ and $\hat{B}$:

$$\hat{A} + \chi^* \hat{B} = \int x(t) f^*(t) \, dt ,$$

$$\chi \hat{A} + \hat{B} = \int x(t) \, g^*(t) \, dt ,$$

where χ is the ambiguity function,

$$\chi = \int f(t) \, g^*(t) \, dt . \qquad (20\text{–}4)$$

In Chapter 2 where signals $f(t)$ and $g(t)$ differ by time and frequency shifts, this function was considered as well as some more general cases.

The solution of the last system of equations defines the most likely estimates of the comples amplitudes of the signals:

$$\hat{A} = \frac{1}{1 - |\chi|^2} \int x(f^* - \chi^* g^*) \, dt ,$$
$$\hat{B} = \frac{1}{1 - |\chi|^2} \int x(g^* - \chi f^*) \, dt . \qquad (20\text{–}5)$$

* This is also a system of complex conjugate equations.

Hence, it is clear that to find these estimates, the input signal $x(t)$ must be processed by two optimal filters matched to the signals $f(t)$ and $g(t)$ and then the outputs are combined in some linear fashion.*

For a more descriptive interpretation of these relationships, let us turn to the vector representation of the signals in the sampling space. The integral of the product of the two functions corresponds to the scalar product of the vectors; in particular, from (20–4),

$$\chi = \mathbf{f}\mathbf{g}^* . \tag{20–4'}$$

In addition, from (20–1),

$$|\mathbf{f}|^2 = \mathbf{f}\mathbf{f}^* = 1 , \quad |\mathbf{g}|^2 = \mathbf{g}\mathbf{g}^* = 1 . \tag{20–1'}$$

The mathematical operations (20–5) resulting in the estimates $\hat{A}$ and $\hat{B}$ have the following vector notation:

$$\begin{aligned} \hat{A} &= \mathbf{x}\tilde{\mathbf{g}}^* , \\ \hat{B} &= \mathbf{x}\tilde{\mathbf{f}}^* , \end{aligned} \tag{20–5'}$$

where

$$\begin{aligned} \tilde{\mathbf{g}} &= \frac{1}{1 - |\chi|^2} (\mathbf{f} - \chi\mathbf{g}) , \\ \tilde{\mathbf{f}} &= \frac{1}{1 - |\chi|^2} (\mathbf{g} - \chi^*\mathbf{f}) . \end{aligned} \tag{20–6}$$

The vectors $\tilde{\mathbf{g}}$ and $\tilde{\mathbf{f}}$ have the following properties:

$$\begin{aligned} \mathbf{g}\tilde{\mathbf{g}}^* &= \mathbf{f}\tilde{\mathbf{f}}^* = 0 , \\ \mathbf{f}\tilde{\mathbf{g}}^* &= \mathbf{g}\tilde{\mathbf{f}}^* = 1 , \\ |\tilde{\mathbf{g}}|^2 &= |\tilde{\mathbf{f}}|^2 = \frac{1}{1 - |\chi|^2} . \end{aligned} \tag{20–7}$$

These relationships can be readily proved using the vector or the equivalent integral transformations. For example, the important *orthogonality property between vectors* $\mathbf{g}$ and $\tilde{\mathbf{g}}^*$ is proved in the following manner:

$$\begin{aligned} \mathbf{g}\tilde{\mathbf{g}}^* &= \frac{1}{1 - |\chi|^2} \mathbf{g}(\mathbf{f}^* - \chi^*\mathbf{g}^*) = \frac{1}{1 - |\chi|^2} (\mathbf{f}^*\mathbf{g} - \chi^* |\mathbf{g}|^2) \\ &= \frac{1}{1 - |\chi|^2} (\chi^* - \chi^*) = 0 . \end{aligned}$$

Relationships (20–4') and (20–1') are used in the above.

* It is emphasized that here the function χ is a numerical parameter. For example, this corresponds to the quantization of the range (range rate, etc.) into small intervals for which χ is a constant complex number for each region.

In vector terminology, (20–5′), the operation for estimating the complex amplitudes, reduces to determining the projection of the signal vector **x** on the vectors $\tilde{\mathbf{g}}$ and $\tilde{\mathbf{f}}$, which are orthogonal to the expected signals **g** and **f**, respectively. This is illustrated in Figure 26. The vectors **f** and **g** define the

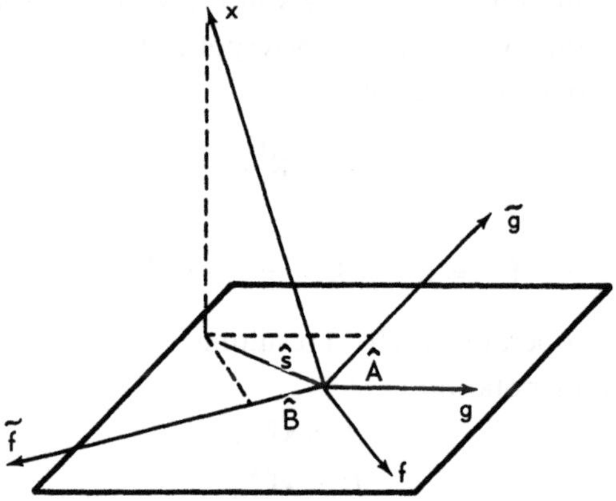

Fig. 26

plane (subspace) of the signals. For arbitrary amplitudes, the signals $\mathbf{s} = A\mathbf{f} + B\mathbf{g}$ fill the subspace. The projection of the vector **x** onto the subspace of the signals yields the most likely estimate of the received signal, $\hat{\mathbf{s}}$. However, in estimating the amplitudes, the *mutual* system of vectors $\tilde{\mathbf{f}}$ and $\tilde{\mathbf{g}}$ and not the basic system of **f** and **g** is used; moreover, the axes of the basic and mutual systems are perpendicular to each other. Such a rule for estimating excludes interaction between signals, since the estimate of the amplitude A is independent of the components of vector **x** coinciding with the direction of **g** (and similarly, for the amplitude B).

Under these conditions, the errors in resolution are caused only by noise, the components of which in the $\tilde{\mathbf{g}}$ or $\tilde{\mathbf{f}}$ directions are large enough to be mistaken for signal (a false alarm). In addition, the noise components may compensate for signal coming from a real target (non-detection). The probabilities of a false alarm and non-detection are denoted by F and Q, respectively. It is noted that, with respect to resolution, both errors have approximately equivalent forms: in the first case, one target is mistaken for two, and in the second case, two targets are mistaken for one.* Therefore, in the theory of resolution, the decisions can be based on the criterion of an ideal observer, *i.e.*,

$$F + Q = \min. \tag{20–8}$$

* The probability of non-detection of both targets and, conversely, the probability of a false alarm with no target present, are neglected in this development. These probabilities equal F^2 and Q^2 and are usually very small.

Since the probability F decreases with an increase in the threshold level, and the probability Q changes inversely, the criterion (20–8) defines the magnitude of the threshold.

In order to calculate the probabilities F and Q, it is necessary to know the distributions of the amplitude estimate $\hat{A}_0 = |\hat{A}|$ (also $\hat{B}_0 = |\hat{B}|$). Substituting $\mathbf{x} = \mathbf{n} + A\mathbf{f} + B\mathbf{g}$ into (20–5′) and considering the (20–7) relationships, we obtain

$$\hat{A} = A + \mathbf{n}\tilde{\mathbf{g}}^* = A_0 \, e^{j\theta} + n_g . \qquad (20\text{–}9)$$

The estimate $\hat{A}$ is a random complex variable which depends on the statistical properties of the signal A (but not on those of the signal B). The last term in (20–9),

$$n_g = \mathbf{n}\tilde{\mathbf{g}}^* = \frac{1}{1 - |\chi|^2} \int n(t) \left[f^*(t) - \chi^* g^*(t)\right] dt ,$$

represents the noise passed by the filter matched to the signal $\tilde{\mathbf{g}}$. The standard deviation of this noise is computed from the following [see the deviation of (18–9)]:

$$2\sigma^2 = |n_g|^2 = |\mathbf{n}\tilde{\mathbf{g}}^*|^2 = |\mathbf{n}|^2 \, |\tilde{\mathbf{g}}|^2 = \frac{2N}{1 - |\chi|^2} . \qquad (20\text{–}10)$$

Here N is the noise density at the input, and the factor 2 is associated with the presence of the imaginary component.

It follows from (20–9) that the resolution phenomenon being considered here can be interpreted as the problem of detecting a signal against a noise background which has random initial phase and unknown amplitude. Therefore, the probability distribution of the amplitude estimate $\hat{A}_0$ can be taken from detection theory (*e.g.*, Helstrom [70], p. 129). However, since this analogy is not quite obvious in some of the details, it is developed in Appendix 3, where advantage is taken of the simplification due to the one-dimensionality of the resolution.

The probability distribution of the amplitude estimate $\hat{A}_0$ is

$$p(\hat{A}_0) = \frac{\hat{A}_0}{\sigma^2} \, e^{-\frac{A_0^2 + \hat{A}_0^2}{2\sigma^2}} \, I_0\left(\frac{A_0 \hat{A}_0}{\sigma^2}\right),$$

where I_0 is the modified Bessel function.

In the absence of signal ($A_0 = 0$), this reduces to the Rayleigh distribution:

$$p_0(\hat{A}_0) = \frac{\hat{A}_0}{\sigma^2} \, e^{-\frac{\hat{A}_0^2}{2\sigma^2}} .$$

The probabilities of false alarm and non-detection are, respectively,

$$F = \int_{l}^{\infty} p_0(\hat{A}_0)\, d\hat{A}_0 = e^{-\frac{l^2}{2\sigma^2}},$$

$$Q = \int_{0}^{l} p(\hat{A}_0)\, d\hat{A}_0 = 1 - \int_{l}^{\infty} \frac{\hat{A}_0}{\sigma^2} e^{-\frac{A_0^2 + \hat{A}_0^2}{2\sigma^2}} I_0\left(\frac{A_0 \hat{A}_0}{\sigma^2}\right) d\hat{A}_0,$$

where l is the threshold level.

For a high-intensity signal [more precisely, for $(A_0 - l)/\sigma \gg 1$], the last expression is simplified substantially and takes the following form:

$$Q \approx \frac{1}{\sqrt{2\pi}} \frac{\sigma}{A_0 - l} e^{-\frac{(A_0 - l)^2}{2\sigma^2}}.$$

Using this approximation, it is not difficult to show that for criterion (20–8) (the one for an ideal observer), the threshold must be established at approximately one-half the signal level [$l \approx (A_0/2)$]. The probability for any kind of error (either false alarm or non-detection) is

$$F + Q \approx e^{-\frac{A_0^2}{8\sigma^2}} = e^{-\frac{E_A(1 - |\chi|^2)}{8N}}. \qquad (20\text{–}11)$$

Here (20–10), (20–1), and the fact that the energy E_A is equal to A_0^2 are taken into account. There is an analogous relationship for the signal B. These results fully reveal the significance of the ambiguity function in determining the resolution of the signal. The closer the parameters of the two signals, the closer $|\chi|^2$ is to unity, and the greater is the probability for errors. If the signals are identical, then $|\chi|^2 = 1$ and it is not possible to resolve the two signals. However, the error probability is also a function of the signal-to-noise ratio. The parameter which defines the reliability of the resolution is

$$\frac{E}{N}(1 - |\chi|^2) = d_{eq}. \qquad (20\text{–}12)$$

Consequently, either by increasing the signal energy or by decreasing the noise density, it is possible to resolve signals which are as close as desired (but nevertheless, different). It must be understood that, strictly speaking, resolution is the capability to distinguish two signals with a *given probability of error* and the parameter d_{eq} is the measure of this capability.

It is also clear that the effect of the signal form on the quality of the resolution is fully accounted for by $|\chi|^2$. In particular, the ranges and range rates of all targets for which

$$|\chi(t, \Omega)|^2 = \text{const},$$

can be equally resolved (assuming the same signal-to-noise ratios). This fact confirms the fitness of estimating the signal form from the ambiguity function. However, a true estimate of resolution is only possible by considering energy relationships.

Thus, the statistical interpretation yields a quantitative confirmation of the qualitative result of the previous section. *The resolution depends substantially on the ambiguity function (equipment function), but by increasing the signal energy there is no limit in the resolution.*

It is noted, in conclusion, that the assumptions of the previous section are not essential for the above result. By not using the approximations and by proceeding from a more general detection criterion the error probability of the resolution is expressed by a more complicated relationship, but nevertheless, it depends only on the parameter d_{eq} defined by (20–12).

[§21] THE AMBIGUITY FUNCTION IN RELATION TO THE MEASURED PARAMETERS

Let us denote by the vector $\boldsymbol{\alpha}$ all the measurable parameters of the signal, *i.e.*, the time of arrival and the Doppler shift $[\boldsymbol{\alpha}=\alpha(t, \Omega)]$, and let us investigate the estimation of these parameters which follow from the principle of the maximum likelihood. From §19, it is necessary to find the maximum of the likelihood function with respect to all parameters (those not subject to measurement, as well as those being measured), *i.e.*, to minimize

$$L_x(\boldsymbol{\alpha}, \theta, A) = \int |x(t') - A\, e^{j\theta} s(t'; \boldsymbol{\alpha})|^2\, dt'$$

with respect to all variables.* Let us begin by minimizing the function with respect to the phase θ. If the signal $s(t; \boldsymbol{\alpha})$ is normalized with respect to energy, we can write, using vector notation,

$$\begin{aligned} L_x &= (\mathbf{x} - A\, e^{j\theta} \mathbf{s}(\boldsymbol{\alpha}))\,(\mathbf{x}^* - A\, e^{-j\theta} \mathbf{s}^*(\boldsymbol{\alpha})) \\ &= |\mathbf{x}|^2 + A^2 - 2A\, \mathrm{Re}\, \mathbf{x}\mathbf{s}^*\, e^{-j\theta}. \end{aligned} \tag{21–1}$$

Only the last term is dependent on θ. Denoting

$$\mathbf{x}\mathbf{s}^* = z = |z|\, e^{j \arg z},$$

we obtain

$$\mathrm{Re}\, \mathbf{x}\mathbf{s}^*\, e^{-j\theta} = |z| \cos(\arg z - \theta).$$

* Here A is understood to be a real variable, *i.e.*, the expected signal amplitude A_0 denotes the amplitude of the received signal.

The above has a maximum when $\theta = \arg z$, and this yields a minimum of (21–1). Consequently,

$$\min_{\theta} L_x = E_x + A^2 - 2A\,|z|\,,$$

where $E_x = |\mathbf{x}|^2$ is the energy of the received signal. The minimum with respect to A is found in the ordinary manner. As a result,

$$\min_{\theta, A} L_x = E_x - |z|^2 = E_x - |\int x(t')\, s^*(t'; \boldsymbol{\alpha})\, dt'|^2 .$$

Thus, for a signal with random phase and amplitude, the problem reduces to maximizing the magnitude of the *correlation integral* (mutual correlation integral),

$$|z(\hat{\boldsymbol{\alpha}})|^2 = |\int x(t')\, s^*(t'; \hat{\boldsymbol{\alpha}})\, dt'|^2 = \max . \qquad (21\text{–}2)$$

The function $z(\boldsymbol{\alpha})$ is formed at the output of the filter matched to the signal $\mathbf{s}(\boldsymbol{\alpha})$. The measurement procedure that results in the parameter estimates from the maximum likelihood criterion can be carried out as follows.

A set of parallel filters matched to the transmitted signal, but with various frequency shifts, are used. After each filter there is a square law detector which separates the envelope and processes the signal according to (21–2). The block diagram of this system was given in §1 (see Fig. 1). In every channel corresponding to the various Doppler shifts, there are many outputs caused by both the noise and target reflections. If the signal level is high, there is one principal peak exceeding all the noise outputs. The maximum peaks of all channels are compared, and the principal maximum is chosen (*i.e.*, the channel number) to obtain the final estimate of the arrival time and Doppler shift.

If there is no noise, such an estimate corresponds to the true target position. In fact, the input signal $x(t)$ can be represented by

$$x(t) = A_0\, e^{j\theta_0} s(t; \boldsymbol{\alpha}_0),$$

where A_0 and θ_0 are the signal amplitude and phase (the unknown quantities), and $\boldsymbol{\alpha}_0 = \alpha(t_0, \Omega_0)$ is the vector representation of the target parameters. Therefore, $|z(\boldsymbol{\alpha})|$ is proportional to the magnitude of the ambiguity function:

$$|z(\boldsymbol{\alpha})| = A_0\, |\mathbf{s}(\boldsymbol{\alpha}_0)\, \mathbf{s}^*(\boldsymbol{\alpha})| = A_0\, |\chi(\boldsymbol{\alpha} - \boldsymbol{\alpha}_0)|\,,$$

which is maximum for $\boldsymbol{\alpha} = \hat{\boldsymbol{\alpha}} = \boldsymbol{\alpha}_0$.

In general, the correlation integral also contains a noise component:

$$z(\boldsymbol{\alpha}) = [A_0\, e^{j\theta_0} \mathbf{s}(\boldsymbol{\alpha}_0) + \mathbf{n}]\, \mathbf{s}^*(\boldsymbol{\alpha}) = A_0\, e^{j\theta_0} \chi(\boldsymbol{\alpha} - \boldsymbol{\alpha}_0) + \mathbf{s}^*(\boldsymbol{\alpha})\, \mathbf{n}\,, \qquad (21\text{–}3)$$

which displaces the position of the maximum and causes measurement errors.

On the basis of (21–3), we have

$$|z(\boldsymbol{\alpha})|^2 = E\,|\chi(\boldsymbol{\alpha} - \boldsymbol{\alpha}_0)|^2 + 2\,\mathrm{Re}\sqrt{E}\,\mathrm{e}^{-j\theta} \times \chi(\boldsymbol{\alpha} - \boldsymbol{\alpha}_0)\,\mathbf{s}^*(\boldsymbol{\alpha})\,\mathbf{n} + |\mathbf{n}|^2, \qquad (21\text{–}4)$$

where $E = A_0^2$ is the energy of the reflected signal.

It is clear from the above that reliable measurements are feasible when the signal is sufficiently strong so that the correlation integral due to signal substantially exceeds the noise outputs. Under these conditions, the last term in (21–4) can be neglected. Using this approximation, we obtain from (21–4),

$$\frac{1}{\sqrt{E}}|z(\boldsymbol{\alpha})| = u(\boldsymbol{\alpha}) = |\chi(\boldsymbol{\alpha} - \boldsymbol{\alpha}_0)| + v(\boldsymbol{\alpha}), \qquad (21\text{–}5)$$

where

$$v(\boldsymbol{\alpha}) = \mathrm{Re}\,\frac{1}{\sqrt{E}}\,e^{-j(\theta_0 - \arg\chi)}\,\mathbf{s}^*(\boldsymbol{\alpha})\,\mathbf{n}. \qquad (21\text{–}6)$$

The estimate of the signal parameters, which corresponds to a correlation integral maximum, can be determined from the system of equations,

$$\left.\begin{aligned} \frac{\partial u}{\partial t} = \frac{\partial|\chi|}{\partial t} + \frac{\partial v}{\partial t}\bigg|_{\substack{t=\hat{t}\\ \Omega=\hat{\Omega}}} &= 0, \\ \frac{\partial u}{\partial \Omega} = \frac{\partial|\chi|}{\partial \Omega} + \frac{\partial v}{\partial \Omega}\bigg|_{\substack{t=\hat{t}\\ \Omega=\hat{\Omega}}} &= 0, \end{aligned}\right\} \qquad (21\text{–}7)$$

obtained by differentiating (21–5). However, this system of equations has many solutions which correspond not only to the principle maximum but also to noise spikes. Let us make the following assumption in order to simplify the investigation.

Since the noise level is low, the expected measurement errors are also small. This means that only the central region of the ambiguity function is of importance. The expansion of the ambiguity function about its center contains only second degree terms and is

$$\chi(t - t_0, \Omega - \Omega_0) = 1 + \tfrac{1}{2}\{h_{11}(t - t_0)^2 + 2h_{12}(t - t_0)(\Omega - \Omega_0) + h_{22}(\Omega - \Omega_0)^2\}. \qquad (21\text{–}8)$$

The factors h_{11}, h_{12}, and h_{22} are the second partial derivatives of the ambiguity function evaluated at its center. The following are the values of these derivatives obtained in §10:

$$h_{11} = \frac{\partial^2 \chi}{\partial t^2} = -\frac{1}{2\pi} \int_{-\infty}^{\infty} \omega^2 |S(\omega)|^2 \, d\omega = -4\pi^2 F^2 ,$$

$$h_{22} = \frac{\partial^2 \chi}{\partial \Omega^2} = -\int_{-\infty}^{\infty} t^2 |s(t)|^2 \, dt = T^2 ,$$

$$h_{12} = \frac{\partial^2 \chi}{\partial t \, \partial \Omega} = -\int_{-\infty}^{\infty} t\varphi'(t) \, |s(t)|^2 \, dt ,$$

where φ is the signal's phase structure. Since these derivatives are real, the function χ is real and coincides with the magnitude $|\chi|$ in this approximation. As a result, the system of equations (21–7) takes the form

$$\left.\begin{aligned} h_{11}(\hat{t} - t_0) + h_{12}(\hat{\Omega} - \Omega_0) + v_t = 0 , \\ h_{12}(\hat{t} - t_0) + h_{22}(\hat{\Omega} - \Omega_0) + v_\Omega = 0 . \end{aligned}\right\} \tag{21–9}$$

From (21–6), v_t and v_Ω are random variables having the form

$$v_t = \mathrm{Re} \, \frac{1}{\sqrt{E}} \, \mathrm{e}^{-j\theta_0} \frac{\partial \mathbf{s}^*}{\partial t} \, \mathbf{n} ,$$

$$v_\Omega = \mathrm{Re} \, \frac{1}{\sqrt{E}} \, \mathrm{e}^{-j\theta_0} \frac{\partial \mathbf{s}^*}{\partial \Omega} \, \mathbf{n} ,$$

where, after carrying out the differentiation, the substitutions $t = \hat{t}$ and $\Omega = \hat{\Omega}$ are made.

Since the variables v_t and v_Ω are formed by linear transformations with noise, they have normal distributions. Their standard deviations are

$$\overline{v_t^2} = \frac{N}{E} \left| \frac{\partial \mathbf{s}^*}{\partial t} \right|^2 = \frac{N}{E} \int \left| \frac{\partial s}{\partial t} \right|^2 dt = \frac{N}{E} \frac{1}{2\pi} \int \omega^2 |S(\omega)|^2 \, d\omega$$
$$= -\frac{N}{E} h_{11} ,$$

and

$$\overline{v_\Omega^2} = -\frac{N}{E} h_{22} .$$

The covariance can be calculated in a similar manner (see Helstrom, [70], p. 236):

$$\overline{v_t v_\Omega} = -\frac{N}{E} h_{12} .$$

Solving the system of equations (21–9) makes it possible to express the measurement errors $\delta t = t - t_0$ and $\delta\Omega = \Omega - \Omega_0$ in terms of the random variables v_t and v_Ω. Furthermore, it is not difficult to find the standard deviations and the covariance of the errors.

$$\begin{aligned}\sigma_t^2 &= \overline{(\hat{t} - t_0)^2} = \frac{N}{E}\frac{h_{22}}{h},\\ \sigma_\Omega^2 &= \overline{(\hat{\Omega} - \Omega_0)^2} = \frac{N}{E}\frac{h_{11}}{h},\\ K &= \overline{(\hat{t} - t_0)(\hat{\Omega} - \Omega_0)} = \frac{N}{E}\frac{h_{12}}{h},\end{aligned} \tag{21–10}$$

where

$$h = h_{11}h_{22} - h_{12}^2 .$$

These expressions for the standard deviations and covariance of the measurement errors are well known (Helstrom [70], p. 227; Fal'kovich [63], p. 132). These results, which are a quantitative estimate of the measurement accuracy, manifest the role of the signal level and its form (spectrum width and duration) in the measurement process. From the engineering viewpoint, the results are completely sufficient.

However, due to the simplifications made here, the form of the ambiguity function depends only on coefficients h_{11}, h_{22}, and h_{12}. In order to emphasize the role of the ambiguity function, we shall attempt to obtain a similar expression for the error probability, which contains the ambiguity function explicitly and is not in the form of the expansion (21–8). However, such an expression cannot be derived on the basis of the approximate theory used above. Some arbitrariness is unavoidable; it is associated with extrapolating the expressions suitable for the small error region over a wider region.

Since the errors δt and $\delta\Omega$ are correlated random variables with normal distributions, we can write the combined probability density of δt and $\delta\Omega$ as a function of the standard deviations and covariance as follows (see Davenport & Root [23], p. 148):

$$p(\delta t, \delta\Omega) = \frac{1}{2\pi\sqrt{\sigma_t^2\sigma_\Omega^2 - K^2}}\, e^{-\frac{\sigma_\Omega^2\delta t^2 - 2K\delta t\delta\Omega + \sigma_t^2\delta\Omega^2}{2(\sigma_t^2\sigma_\Omega^2 - K^2)}} .$$

By using (21–10), this becomes

$$p(\delta t, \delta\Omega) = \frac{\sqrt{h}}{2\pi}\frac{E}{N}\, e^{\frac{E}{2N}(h_{11}\delta t^2 + 2h_{12}\delta t\delta\Omega + h_{22}\delta\Omega^2)} . \tag{21–11}$$

Comparing this relationship with (21–8), it can be assumed that the combined probability density can be expressed in closed form by

$$p(\delta t, \delta\Omega) = \gamma\, e^{-\frac{E}{2N}(1-|\chi(\delta t, \delta\Omega)|^2)}, \tag{21–12}$$

where the coefficient γ is determined from the ordinary normalizing condition and depends on the form of the ambiguity function.

Limiting oneself to the second-order terms in the expansion of $|\chi(\delta t, \delta\Omega)|^2$, it is readily seen that expressions (21–11) and (21–12) are the same.

Thus, at least for small measurement errors, the combined probability density is a function of the same parameter:

$$d_{eq} = \frac{E}{N}(1 - |\chi|^2). \tag{21–13}$$

Therefore, the errors in signal resolution depend on d_{eq} [*cf.* (20–12)].*

Obviously, the basic deductions of the previous section, *i.e.*, that the measurement accuracy depends on the form of the ambiguity function, and that by increasing the signal energy there is no limit to the accuracy of the simultaneous measurement of frequency and time, pertain fully here.

Let us return to relationship (21–11) which characterizes the combined probability of small errors in time and frequency. It can be shown (*e.g.*, see Dulevich, *et al.* [26], p. 515), that the probability of the total error lying inside a "scatter ellipse" of the form

$$h_{11}\tau^2 + 2h_{12}\tau\Omega + h_{22}\Omega^2 = -c^2 \tag{21–14}$$

is given by

$$P_c = 1 - e^{-\frac{E}{2N}c^2}.$$

The area of ellipse (21–14) is proportional to c^2 and has a weak dependence on the signal parameters. Therefore, for a fixed value of the probability P_c, the area of the ellipse, which is a convenient measure of the accuracy when simultaneous measurements of time and frequency are made, is dependent

* There are some indirect indications that (21–12) and particularly (21–13) have a wider region of applicability. The following results are cited. In order for the estimates $\hat{t}$ and $\hat{\Omega}$ to take on values other than the true ones, it is necessary (but not sufficient) that the vatue of the correlation integral evaluated at these estimates, $\hat{\alpha} = \alpha(\hat{t}, \hat{\Omega})$, be greater than the integral evaluated at the true target position, $\alpha_0 = \alpha(t_0, \Omega_0)$. Using (21–4), it can be shown that the probability of such an event is defined by the integral

$$\sqrt{\frac{d_{eq}}{8\pi}} \int_1^\infty e^{-\frac{d_{eq}x^2}{8}}\, dx$$

and consequently depends only on the parameter d_{eq}. This expression is valid for all values of δt and $\delta\Omega$ (and is not constrained to small errors).

mainly on the signal-to-noise ratio:

$$\Delta = k\frac{N}{E}, \tag{21–15}$$

where the coefficient k depends on the probability P_c. An analogous result will be obtained in the next section by another method based on some considerations of a more general physical nature. It should also be noted that instead of using (21–11), which is an approximate expression of the combined probability, a more general expression [such as (21–12)] makes it possible to obtain an estimate of the scatter ellipse's effective area. The Cramér-Rao inequality yields such an estimate (see Cramér [35], p. 480; Middleton [48], p. 943).

[§22] RADAR OBSERVATIONS AND MEASUREMENTS IN RELATION TO GENERAL PHYSICS

Some of the problems associated with estimating the accuracy of radar measurements can be interpreted from the viewpoint of the general theory of physical experiments. Such a theory was developed by Brillouin ([5], pp. 184–243).

The basic premise of this theory is the assertion that any measurement is associated with an energy transfer which is sufficiently large to exceed the thermal noise background present in any measuring device. The energy expended on the observation is partially converted into heat, thus increasing the noise fluctuations. Moreover, the larger the amount of information measured, or the greater the required accuracy, the larger the energy expenditure and thus the higher the level of the fluctuations.

The quantitative relationship between information and energy loss is given by the negative-entropy principle of information which is a generalization of the second law of thermodynamics. This principle can be written as

$$dI \leqslant d\mathscr{E},$$

where $d\mathscr{E} = dQ/T$ = change in the entropy of a system;
T = temperature (°K), defining the fluctuation level;
dQ = increase of heat associated with the measurement;
dI = increment of information obtained as a result of the measurement.

In general, not all the energy expended on the measurement, dE, is converted into heat, *i.e.*, $dQ < dE$ and $dI \leqslant dE/T$. In integral form we also have

$$\Delta I = I_2 - I_1 \leqslant \int_{Q_1}^{Q_2} \frac{dQ}{T} \leqslant \int_{E_1}^{E_2} \frac{dE}{T}, \tag{22–1}$$

where the subscripts 1 and 2 refer to the initial and final states of the system, *i.e.*, before and after the measurement.

If it is known prior to the measurement that the independent variable is in a given interval L, and as a result of the measurement the position of this independent variable has an error to within ΔL, then

$$\Delta I = k \ln \frac{L}{\Delta L}. \tag{22-2}$$

The coefficient k depends on the choice of the information units. For the thermodynamic units used in the following, $k=1.38\cdot 10^{-16}$ *erg/deg* (Boltzmann constant).

The negative entropy principle can be described as follows. Consider an isolated thermodynamic system containing an energy source, observation object, measuring device, medium in which the measurements are made, and all the auxiliary equipment participating in the measurement process. If there is no energy flow with the outside, then all the energy accumulated in the system will be gradually converted into heat, *i.e.*, into random fluctuations. When every element of volume contains, on the average, equal amounts of energy, a thermal equilibrium will exist.

The energy of a system at equilibrium cannot be concentrated, even partially, in particular volume elements or have some non-uniform distribution (*e.g.*, have some non-uniform distribution in time or frequency). The fluctuations of a system at equilibrium have the nature of stationary white noise and are identical at every point. Entropy, being a measure of a system's "randomness", is at a maximum in the equilibrium state when the energy is uniformly distributed in space with respect to time, frequency, etc.

The second law of thermodynamics asserts that any process associated with the transfer and conversion of energy in a non-equilibrium system brings the system closer to its equilibrium state and thus increases the total entropy. The energy is transferred from a more organized, non-equilibrium form to a less organized, equilibrium form; energy transfer in the reverse direction is not possible. Since every measurement is also associated with an energy transformation, it follows that the deficiency of entropy, *i.e.*, the difference between the entropy of the system at its present state and the entropy of the system at equilibrium, is a measure of the capability to perform the measurement. The concept of negentropy (negative entropy is introduced for the purpose of convenience) and the capability to make a measurement is characterized by the reserve of negentropy in the system.

Usually the measuring procedure can be divided into two steps. To begin with, negentropy is related to the energy sources (batteries, etc.). During the measurement, a portion of the energy is converted into a signal which transmits information on the observation. This can be interpreted as a transfer from negentropy into information. Then the signals are picked up by the receiver and the energy is used as, for example, to obtain a marking on the screen of an oscilloscope. Moreover, the information is then again converted into negentropy since the appearance of a luminous spot on the screen results in a more non-uniform energy distribution than before when the entire screen had identical brightness. However, both transformations (negentropy into information and vice versa) are associated with losses. Part of the negentropy is lost during the first transformation and part of the information is lost during the second transformation.

A thorough analysis of various physical experiments shows that the larger the negentropy loss, the higher the measurement accuracy; the negentropy principle of information establishes the minimum cost of any measurement or observation.

When applied to communication theory, the loss of information and negentropy has a somewhat broader sense than presented above. In communication systems, the losses in the transmission channel are frequently neglected, *i.e.*, we neglect that part of the signal energy which is converted into heat and increases the system entropy. Instead, it is assumed that the noise and signal at the input are independent. In such a model, the noise signal is in no way related to the energy losses of the signal. However, even if the initial energy is "pure" signal without any noise, the extraction of the information results in the generation of the heat with part of the information being destroyed and the signal is irreversibly distorted. At the output of a channel, only a portion of the signal energy can yield information with the rest of the energy being "lost" to the observer. These losses in the signal energy are equivalent to increasing the energy of the noise fluctuations, since part of the signal is irreversibly converted into noise and cannot be distinguished from the equipment noise. If at the input of a channel there is a signal with energy E_1 and the noise has energy Q_1, the total energy at the output of this channel equals the sum of these two factors. Therefore, $E_1+Q_1=E_2+Q_2$; since the signal energy at the output $E_2<E_1$, the equivalent noise at the output $Q_2>Q_1$. Such a change in noise energy

$$\Delta Q = Q_2 - Q_1 = E_1 - E_2 \tag{22–3}$$

further restricts relationship (22–1).*

* It was shown by Brillouin that such an interpretation of the mechanism for information loss makes it possible to relate the negentropy principle with the known Shannon equation for the capacity of a channel with noise. Eq. (22–4), which will be given subsequently, is equivalent to the Shannon equation.

Let us assume that at the input of a receiver we have a signal with energy E and noise with density N. The temperature T_1, which describes the initial level of the fluctuations, is related to the noise density by the Nyquist equation:*

$$N = kT_1,$$

where k is the Boltzmann constant.

The energy of the noise arriving at the input of a receiver with bandwidth Δf during the observation time τ is

$$Q_1 = N\Delta f\tau = kT_1\Delta f\tau.$$

As a result of the interaction between the signal and noise, part of the information contained in the signal is destroyed and the effective noise temperature is increased at the expense of causing signal distortions:

$$T = T_1 + \frac{\xi E}{k\Delta f\tau},$$

where $\xi \leqslant 1$ is the ratio of the signal energy undergoing irreversible conversion.

Furthermore, the negentropy principle of information (22–1) leads to the following relationship:

$$\Delta I \leqslant \int_{Q_1}^{Q_2} \frac{dQ}{T} = \int_0^{\xi E} \frac{d(\xi E)}{\dfrac{N}{k} + \dfrac{\xi E}{k\Delta f\tau}} = k\Delta f\tau \ln\left(1 + \frac{\xi E}{N\Delta f\tau}\right).$$

Since $\xi \leqslant 1$, we have

$$\Delta I \leqslant k\Delta f\tau \ln\left(1 + \frac{E}{N\Delta f\tau}\right). \tag{22–4}$$

Transferring over to the concepts of the radar measurement of range and range rate, we assume that the reflected signal is "already detected". This means that the time of arrival of the signal lies within the observational interval τ, and its frequency lies within the receiver bandwidth Δf. Consequently, a region in t, Ω-plane where the return signal must fall is known prior to the measurement; one of the points in this region corresponds to the true parameters of the target. After the measurement, the target position on this plane is determined more accurately: the measurement error lies within a certain ellipse, the area of which is Δ. It follows from (22–2) that

* See Brillouin ([5], p. 254) concerning the difference between this expression and the frequently encountered relationship $N = 4kT$.

the information obtained during the measurement is

$$\Delta I = k \ln \frac{m}{\Delta},$$

where $m = \tau f > \Delta$. Therefore, (22–4) yields

$$\ln \frac{m}{\Delta} \leqslant m \ln\left(1 + \frac{E}{Nm}\right)$$

or

$$\Delta \geqslant m\left(1 + \frac{E}{Nm}\right)^{-m}. \tag{22–5}$$

The negative-entropy principle of information characterizes the potentialities of the most economical measurement with the total energy expenditures minimized. Therefore, it must be assumed that the observation time, τ, is approximately equal to the signal duration, and that the equipment bandwidth, Δf, equals approximately the signal spectrum. Consequently, the factor m in (22–5) corresponds to the compression ratio; moreover, $m \geqslant 1$. If $m = 1$, we have

$$\Delta \geqslant \frac{1}{1 + \frac{E}{N}} \approx \frac{N}{E}.$$

Obviously, this result agrees with the accuracy of the estimate obtained previously [*cf.*, (21–15)]. However, it must be kept in mind that the measure of accuracy, from the information viewpoint, differs substantially from the quadratic measure used in statistical theory. It is natural, therefore, that in the general case (for $m \neq 1$) there is only a qualitative correspondence: the accuracy obtained when simultaneously measuring both range and range rate depends directly on the signal-to-noise ratio. The above deduction is mainly of interest because such a result essentially follows from a fundamental physical principle: the second law of thermodynamics. Thus, the strong dependence of the radar measurement accuracy on the signal-to-noise ratio acquires an additional, and exceedingly forceful, confirmation.

4

Synthesis of Signals Using the Ambiguity Function

[§23] SOME REMARKS ON SIGNAL SYNTHESIS

The following summarizes the main results of Chapter 2. For every form of received signal, $s(t)$, there is a corresponding ambiguity function,

$$\chi(t, \Omega) = \int_{-\infty}^{\infty} s\left(t' + \frac{t}{2}\right) s^*\left(t' - \frac{t}{2}\right) e^{j\Omega t'} dt'$$

$$= \frac{1}{2\pi} \int_{-\infty}^{\infty} S\left(\omega - \frac{\Omega}{2}\right) S^*\left(\omega + \frac{\Omega}{2}\right) e^{j\omega t} d\omega, \qquad (23\text{–}1)$$

which has the following fundamental properties.

Normalizing the signal with respect to energy ($E = 1$), the ambiguity function has a maximum at $t = \Omega = 0$; moreover,

$$|\chi|_{max} = \chi(0, 0) = 1. \qquad (23\text{–}2)$$

Also, the ambiguity function "volume" does not depend on the signal form:

$$\iint_{-\infty}^{\infty} |\chi(t, \Omega)|^2 \, dt \, d\Omega = 2\pi. \qquad (23\text{–}3)$$

The function $\chi(t, \Omega)$ is the normalized voltage at the output of the optimal receiver which ensures the maximum signal-to-noise ratio. The function's dependence on Ω characterizes the frequency change caused by the Doppler shifts of signal reflected from moving targets. The magnitude $|\chi(t, \Omega)|$ defines the signal properties from the viewpoint of accuracy and resolution of the measured ranges and range rates. The more compressed the ambiguity function volume is in the t or Ω direction, the greater the corresponding accuracy and resolution. The invariance of the volume expressed by (23–3) shows that it is not possible to increase arbitrarily the accuracy and resolution of both the range and range rate simultaneously by selecting the signal form. This statement is the uncertainty principle in radar.

In Chapters 2 and 3, some of the refinements of this uncertainty principle were discussed. It was shown that in estimating the accuracy and resolution, we cannot neglect the effect of noise and other interferences. In the absence of noise (or for a reflected signal with unlimited energy), there is no limitation to the accuracy and resolution. Under these circumstances, it is possible, in principle, to obtain a resolution and accuracy as high as desired for any signal form. Therefore, some caution should be observed when interpreting the radar uncertainty principle.

However, by disregarding the effect of interferences, by assuming the invariance of the energy relationships, and by considering only the dependence of the accuracy and resolution on the *signal form*, the ambiguity function proves to be very useful. With the aid of this function, it is possible to make a relative comparison of various signal forms from the viewpoint of accuracy and resolution. It was shown in the examples of §9, that by using the ambiguity function, it is easy to establish in which regions of the t, Ω-plane the target resolution is best (and worst) and for which values of ranges and range rates the measurement is more accurate. In some cases it is also clear how the resolution depends on the signal structure. Thus, in spite of the fact that the accuracy and resolution depend on many factors (in particular on the method of signal processing), the ambiguity function can be considered as dependent on only the signal form, and thus it characterizes the most important parameters of the radar.

Naturally there arises the general question whether a signal can be synthesized from the ambiguity function, *i.e.*, the problem of determining the signal $s(t)$ which realizes the desired ambiguity function $\chi(t, \Omega)$. This exceedingly complicated problem is the compromise of a number of separate interrelated problems, each of which has an independent practical significance. We will point out here the separate problems:

(a) determine the class of functions which are the exact ambiguity functions of arbitrary signals;

(b) determine the sub-clases of ambiguity functions for the various signal structures (FM pulses, pulse-code modulation, etc.) and also find the signals which produce these ambiguity functions;

(c) calculate the signal whose ambiguity function approximates sufficiently well (in one sense or another) the desired signal not belonging to the class of realizable functions;.

(d) investigate the optimal ambiguity function (for a given sub-class) and the corresponding signal which best satisfies a given set of requirements (*e.g.*, the synthesis of an FM signal to obtain the best time of arrival resolution with no requirement for frequency resolution).

It is pointed out that a majority of these problems have not been fully solved yet. Digressing from the possibility of encountering these problems in practice, we still do not know how to synthesize a signal from an ambiguity function having the desired properties. To illustrate the difficulties arising here, we shall consider briefly a direct and obvious method of synthesis.

Let us assume that the desired ambiguity function $\chi(t, \Omega)$ belongs to a class of realizable functions, *i.e.*, there exists a signal $s(t)$ which is related to the function $\chi(t, \Omega)$ by (23–1). In order to find this signal, we can use the fact that the top expression of (23–1) is the Fourier integral with respect to the variable t'. Therefore, due to the relationship of Fourier transform pairs, we can write

$$s\left(t' + \frac{t}{2}\right) s^*\left(t' - \frac{t}{2}\right) = \frac{1}{2\pi} \int_{-\infty}^{\infty} \chi(t, \Omega)\, e^{-j\Omega t'}\, d\Omega\,. \tag{23–4}$$

Using the variable substitutions

$$t' + \frac{t}{2} = u \quad \text{and} \quad t' - \frac{t}{2} = v\,,$$

we obtain

$$\frac{1}{2\pi} \int_{-\infty}^{\infty} \chi(u - v, \Omega)\, e^{-j\frac{u+v}{2}\Omega}\, d\Omega = s(u)\, s^*(v)\,. \tag{23–5}$$

It follows from the previous discussion that (23–5) is a necessary condition for realizing the ambiguity function $\chi(t, \Omega)$. It can be readily shown that (23–5) is also a sufficient condition. In fact, if for a given $\chi(t, \Omega)$ (Eq. 23–5) is satisfied, then the reverse substitution of variables leads to relationship (23–4), after which the Fourier transform yields the initial (23–1) relationship.

Thus, we have a necessary and sufficient condition for realizing the given ambiguity function. An analogous condition containing the signal spectrum is obtained using the second expression of (23–1). Furthermore, by setting $v = 0$ in (23–5) and changing the notation, we have

$$\frac{1}{2\pi} \int_{-\infty}^{\infty} \chi(t, \Omega)\, e^{-j\frac{\Omega t}{2}}\, d\Omega = s^*(0)\, s(t)\,. \tag{23–6}$$

This relationship allows us to find the desired signal $s(t)$ from the given, realizable ambiguity function. The additional constant multiplier $s^*(0)$ is, obviously, of no significance.

Let us consider an example. It was shown in §9 that for a Gaussian pulse with linear FM, the ambiguity function is as follows:

$$\chi(t, \Omega) = e^{-\frac{1}{2}\left[m^2 \frac{t^2}{T^2} + (\gamma T^2)\,\Omega t + \frac{\Omega^2 T^2}{4}\right]} e^{j\omega_0 t},$$

where $m^2 = 1 + \gamma^2 T^4$, and 2γ is the frequency rate of change.

On the basis of (23–6), we find, with the constant multiplier neglected,

$$s(t) = \frac{1}{2\pi} e^{-\frac{1}{2}m^2 \frac{t^2}{T^2} + j\omega_0 t} \int_{-\infty}^{\infty} e^{-\frac{1}{2}\left[\frac{\Omega^2 T^2}{4} + (\gamma T^2 - j)\,\Omega t\right]} d\Omega$$

$$= \sqrt{\frac{2}{\pi}} \frac{1}{T} e^{-\frac{t^2}{T^2}} e^{j(\omega_0 t + \gamma t^2)}.$$

Thus, the input FM signal is obtained, and it may seem that such a synthesis method satisfactorily solves the problem. However, in practice this solution is of almost no significance for the following reasons. In the first place, it is useful only for realizable ambiguity functions; but as a rule, it is necessary to seek an approximation of the ambiguity function because situations in which the function is realizable are extremely rare. In the second place, to find a solution, it is necessary to know not only the magnitude but also the phase of the ambiguity function; however, from practical considerations, only the magnitude of the function is known with the phase being unknown.* Therefore, an exact solution cannot be obtained even for those cases where the magnitude of the chosen ambiguity function is realizable.

Proceeding to the synthesis problem, the phase must be arbitrarily assigned. Naturally, this leads to solutions which are not necessarily optimal. In order to verify the quality of the approximation, it is necessary to perform calculations for various phase functions and to estimate the errors for each case. Unfortunately, such calculations are very complicated and their effectiveness is not known beforehand. The synthesis problem becomes particularly difficult when it is required to realize a particular ambiguity function form over the entire t, Ω-plane. Frequently it is sufficient, for practical purposes, to solve a less general problem: synthesize a signal which realizes the needed resolution only of range, or only of range rate, or finally of both range and range rate, but in such a manner that the range measurements are made for a stationary target and the range rate of targets at a fixed range. In these cases the ambiguity function, $\chi(t, \Omega)$, is not given over the entire plane but only for the principal cross sections: for $\chi(t, 0)$, $\chi(0, \Omega)$, or both

* The reader is reminded, as was shown in Chapter 3, that it is only the magnitude of the ambiguity function which affects the accuracy and resolution of the range and range rate measurement.

$\chi(t, 0)$ and $\chi(0, \Omega)$. Similarly, it is noted that in the antenna pattern synthesis problem, one usually considers only the pattern in the vertical and horizontal planes and not the entire spatial pattern for all angles. In the following sections we shall first consider the special problems mentioned above and not return to the general formulation until §35.

[§24] CLASSES OF SIGNALS AND AUTOCORRELATION FUNCTIONS—STATEMENT OF THE SYNTHESIS PROBLEM

The most important, and at the same time the simplest, of the problems being considered is that of synthesizing a signal from the autocorrelation function

$$R(t) \equiv \chi(t, 0) = \int_{-\infty}^{\infty} s\left(t' + \frac{t}{2}\right) s^*\left(t' - \frac{t}{2}\right) dt$$

$$= \frac{1}{2\pi} \int_{-\infty}^{\infty} |S(\omega)|^2 \, e^{j\omega t} \, d\Omega \,. \tag{24–1}$$

This is the classical problem of synthesizing a signal to observe stationary (or slowly moving) targets when high range resolution is required. As a rule, the central peak of the autocorrelation function must be as narrow as possible with the sidelobes in the time direction being distributed more or less uniformly. It is also possible to synthesize the signal function from any arbitrary autocorrelation function $R(t)$.

Proceeding to the synthesis, it is necessary to limit somehow the class of admissible signals. This class must be sufficiently wide so as not to limit substantially the possible solutions, and on the other hand it must be instrumental in constructing a closed synthesis theory. Unless otherwise specified, the solutions sought in the following will be limited to signals with bounded spectra, *i.e.*, it is assumed that the spectrum $S(\omega)$ differs from zero only in the finite frequency band*

$$\omega_0 - \frac{\Omega_m}{2} \leqslant \omega \leqslant \omega_0 + \frac{\Omega_m}{2} \,. \tag{24–2}$$

This assumption must be given more detailed consideration. Strictly speaking, the requirement of finite spectrum width is incompatible with pulses having a finite duration; thus, if we assume that the spectrum is bounded, we can only consider signals which are not bounded in time. How-

* It is assumed that the complex signal is formed by complementing the real part by using the Hilbert transform, and thus negative frequencies are absent from the spectrum $S(\omega)$.

ever, a major portion of the signal energy (even for signals with finite duration) is concentrated in a specific frequency band, which makes it possible to approximate the bounded spectrum condition. In particular, we have seen that by using the stationary phase method the spectra of a wide-band FM pulse (bounded in time) can be considered as approximately bounded by the limits of the instantaneous frequency change (see Appendix 1). Thus, at least in those cases where the stationary phase method is applicable to the spectrum calculation, it can be assumed that the spectrum is bounded. In general, the justification of such an assumption depends on the known properties of the signal function.

Any arbitrary function $s(t)$ with finite energy can be approximated by the function $g(t)$, the spectrum of which is bounded by the frequency limits (24–2). The integrated squared difference between the signal and the approximated function,

$$\varepsilon = \int_{-\infty}^{\infty} |s(t) - g(t)|^2 \, dt,$$

is numerically equal to the energy outside the assumed frequency band, *i.e.*,

$$2\pi\varepsilon_{\min} = \int_{-\infty}^{\omega_0 - \frac{\Omega_m}{2}} |S(\omega)|^2 \, d\omega + \int_{\omega_0 + \frac{\Omega_m}{2}}^{\infty} |S(\omega)|^2 \, d\omega .$$

This means that for a sufficiently wide frequency band, the error can be made arbitrarily small (see Khurgin & Yakovlev [71], p. 148).

In addition, it is known that any arbitrary continuous signal $s(t)$ can be estimated by a linear approximation in any finite time interval by a function with a bounded spectrum, *i.e.*, the difference $|s(t) - g(t)|$ at any instant of time (within this interval) is smaller than a given quantity (Khurgin & Yakovlev [71], p. 135; Slepian & Pollak [41], pt. 1). A linear criterion for the approximation leads, generally speaking, to another approximating function. Also, this approximation is valid for any arbitrary time interval and for any narrow frequency band Ω_m. In this case, the integrated energy of the approximating signal, $g(t)$, increases and may substantially exceed the energy of the actual signal, $s(t)$, being approximated. However, by widening the frequency band, the energy of the signal function $g(t)$ approaches the energy of the actual signal, and both approximations (on the basis of the quadratic and linear criteria) approach one another. This means that it is possible to find a signal function with a bounded spectrum that differs from any given arbitrary signal (with finite energy) by a negligibly small amount.

A signal with a bounded spectrum can be represented as follows:

$$s(t) = \frac{1}{2\pi} \int_{\omega_0 - \frac{\Omega_m}{2}}^{\omega_0 + \frac{\Omega_m}{2}} S(\omega)\, e^{j\omega t}\, d\omega\,;$$

by translating the origin of the coordinate axes to the middle of the frequency band and by dropping the unessential multiplier, it becomes

$$s(t) = \frac{1}{2\pi} \int_{-\frac{\Omega_m}{2}}^{\frac{\Omega_m}{2}} S(\omega)\, e^{j\omega t}\, d\omega\,. \tag{24–3}$$

Using (24–1), the following analogous expression of the autocorrelation function is obtained:

$$R(t) = \frac{1}{2\pi} \int_{-\frac{\Omega_m}{2}}^{\frac{\Omega_m}{2}} |S(\omega)|^2\, e^{j\omega t}\, d\omega\,. \tag{24–4}$$

For these equations, it is convenient to transfer over to dimensionless variables. Let us introduce the coordinates ξ and η which are proportional to the frequency and time, and which are defined by the following relationships:

$$\xi = \frac{2\omega}{\Omega_m}, \quad \eta = \frac{t\Omega_m}{2}. \tag{24–5}$$

Substituting the above in (24–3) and (24–4) yields (again the unessential constant multipliers are dropped)

$$s(\eta) = \frac{1}{2\pi} \int_{-1}^{1} S(\xi)\, e^{j\eta\xi}\, d\xi\,, \tag{24–6}$$

$$R(\eta) = \frac{1}{2\pi} \int_{-1}^{1} |S(\xi)|^2\, e^{j\eta\xi}\, d\xi\,. \tag{24–7}$$

In the above, the additional normalization condition

$$R(0) = \frac{1}{2\pi} \int_{-1}^{1} |S(\xi)|^2\, d\xi = 1 \tag{24–8}$$

is also used.

In using the dimensionless variables ξ and η, we have essentially introduces universal units for measuring the spectrum and the autocorrelation function. This makes it possible to consider the synthesis problem from only the viewpoint of the signal form, while disregarding its time duration and spectrum width. After the synthesis, it is possible, of course, to convert to the actual parameters by selecting the necessary spectrum together with a corresponding time scale.

We shall now review the basic properties of functions having bounded spectra. It can readily be seen that the function $f(\eta)$, defined as

$$f(\eta) = \frac{1}{2\pi} \int_{-\alpha}^{\alpha} F(\xi)\, e^{j\eta\xi}\, d\xi, \tag{24–9}$$

is an analytical function of the complex variable $\eta = \sigma + j\gamma$. Applying the Bunyakovskii-Schwartz inequality, we obtain the following upper limit for (24–9):

$$|f(\eta)|^2 \leqslant \frac{1}{2\pi} \int_{-\alpha}^{\alpha} |F(\xi)|^2\, d\xi \int_{-\alpha}^{\alpha} |e^{j(\sigma + j\gamma)\xi}|^2\, d\xi$$
$$\leqslant \frac{\alpha}{\pi} |F_{\max}|^2 \int_{-\alpha}^{\alpha} e^{-2\gamma\xi}\, d\xi = \frac{\alpha}{\pi\gamma} |F_{\max}|^2 \sinh 2\gamma\alpha\,.$$

Therefore, for large values of γ,

$$|f(\eta)| < C\, e^{(\alpha+\varepsilon)\gamma} \leqslant C\, e^{(\alpha+\varepsilon)|\eta|}, \tag{24–10}$$

where C is a constant and ε is an arbitrarily small positive number.

This inequality specifies the upper limit of $f(\eta)$ for large complex values of η. The functions of the complex variable η which satisfy (24–10) are called the *integral functions of the finite exponent* α, and these functions are customarily referred to as the W_α class. Thus, we have shown that the signals $s(\eta)$ and their autocorrelation functions $R(\eta)$ belong to the W_1 class. The class differences between $S(\xi)$ and $R(\xi)$ are due to the fact that the spectral density $S(\xi)$ in (24–6) can be complex for real ξ, and the factor $|S(\xi)|^2$ under the integral in (24–7) is always real and positive. Thus, there are fewer classes of the autocorrelation functions than signals having bounded spectra.

Of importance to the synthesis problem is the theorem we have just proved, *i.e.*, the *Wiener-Paley theorem* (Khurgin & Yakovlev [71], p. 82): *any function having a finite exponent* α, *has a bounded spectrum and can be represented by* (24–9).

It is a simple matter to convince oneself that the function $f(\eta)$ belongs to the W_1 class. For large complex values of η, it suffices to verify that condition (24–10) is satisfied for $\alpha = 1$. Unfortunately, it is not possible to find an equally simple criterion which makes it possible to establish that the function's spectrum is positive.* To verify this, it is necessary to determine the spectrum by direct calculation.

Thus, we can assume that signals belong to the W_1 class, *i.e.*, they are integral functions of the first degree; autocorrelation functions belong to a subclass of the W_1 class which contains all functions having positive spectra. This class is denoted by the symbol W_1^+.

It can readily be seen that the synthesis problem cannot be solved uniquely from the given autocorrelation function

$$R(\eta) = \frac{1}{2\pi} \int_{-1}^{1} |S(\xi)|^2 \, e^{j\eta\xi} \, d\xi .$$

Even if the function $R(\eta)$ is realizable (*i.e.*, belongs to the W_1^+ class), it is only possible to determine the power spectrum of the signal $|S(\xi)|^2$ and not the spectral function $S(\xi)$ itself. All signals, the spectra of which differ only in phase, have the same autocorrelation function.** Therefore, the synthesis procedure being considered here can be divided into two steps:

(1) the power spectrum $|S(\xi)|^2$ is determined from the given autocorrelation function $R(\eta)$;

(2) from the determined $|S(\xi)|^2$, one of the signals having such a power spectrum is derived.

Here we shall only discuss the first part of the problem, *i.e.*, the determination of the power spectrum. This step will be called the *synthesis of the autocorrelation function*. It is not difficult to conclude from the previous

* It is possible to find the condition for which the spectrum of the function $f(\eta) = f_1(\eta) + jf_2(\eta)$ is real. A necessary and sufficient condition for this purpose is that the real component, $f_1(\eta)$, be even and that the imaginary component, $f_2(\eta)$, be odd. In our case, such a condition is necessary but not sufficient since the factor $|S(\xi)|^2$ is not only a real but also positive (more accurately, non-negative). It can also be shown that the function $f(\eta)$ has a positive spectrum if and only if it can be represented in the form of the convolution

$$f(\eta) = \int_{-\infty}^{\infty} s(\eta') \, s^*(\eta' + \eta) \, d\eta',$$

i.e., if it represents the autocorrelation function of some signal. However, this condition is of no advantage from the viewpoint of the synthesis, since the verification of the realizability reduces to determining the sought-after signal.

** Such an ambiguity imposes additional requirements for signal synthesis which are of great practical importance. For example, it will be shown subsequently that it is possible to require an FM signal with a rectangular envelope to have a given power spectrum (see §31).

discussion that in order to find the autocorrelation function

$$R(\eta) = R_1(\eta) + jR_2(\eta) = |R(\eta)|\, e^{j\theta(\eta)},$$

it is necessary and sufficient to satisfy the following set of conditions:

$$|R(\eta)| \leqslant C\, e^{(1+\varepsilon)|\eta|}, \tag{24–11a}$$

$$|R(\eta)| = |R(-\eta)|, \quad \theta(\eta) = -\theta(-\eta), \tag{24–11b}$$

$$F(\xi) = \int_{-\infty}^{\infty} R(\eta)\, e^{-j\xi\eta}\, d\eta \geqslant 0. \tag{24–11c}$$

The first condition (which must be satisfied over the entire complex plane of η where $|\eta| \gg 1$ and for any $\varepsilon > 0$) proves that $R(\eta)$ belongs to the W_1 class. The fulfillment of the second and third conditions (on the real axis) guarantees that the spectrum $F(\xi)$ is a real and positive function. If these conditions are satisfied, then the power spectrum is determined in the obvious manner:

$$|S(\xi)|^2 = F(\xi) = \int_{-\infty}^{\infty} R(\eta)\, e^{-j\xi\eta}\, d\eta; \tag{24–12}$$

moreover, from the Wiener-Paley theorem, this spectrum differs from zero in the interval $-1 \leqslant \xi \leqslant 1$. Unfortunately, the methods having any practical value are much more complicated than those we have discussed. There are three reasons for this situation. First, the given autocorrelation functions usually do not result in a realizable signal function. For example, the best $R(\eta)$, from the viewpoint of resolution, has a table-top shape and this form is generally not an analytical function. Second, due to practical considerations, only the magnitude of the autocorrelation function is known; the phase structure of $R(\eta)$ is of no importance in radar. The characteristic features due to these two items on the signal synthesis are considered in §25 and §26.

The third reason is associated with the difference between the W_1 and the W_1^+ classes. The given autocorrelation function $R(\eta)$ can be approximated by a function having a bounded spectrum and which belongs to the W_1 class. Systemized methods for making such an approximation have been discussed in the literature (*e.g.*, see §26). However, the fact that the approximated autocorrelation function belongs to the W_1^+ class [condition (24–11c)] can be verified only after calculating the spectrum $F(\xi)$, *i.e.*, it can only be determined in the last step of the solution. If condition (24–11c) is not satisfied, then the synthesis must be recycled from the beginning by changing, for example, the given autocorrelation function. Moreover, there is no certainty that this new attempt can be brought to a successful conclusion for the same reason.

In practice, the synthesis is simplified because one usually does not start with an arbitrary autocorrelation function $R(\eta)$ but a function with as short a duration as possible. For example, let $f(\eta)$ belonging to the W_1 class [*i.e.*, $f(\eta)$ satisfies condition (24–11a)] be given as the autocorrelation function to realize the signal. The evaluation of (24–12) yields the spectrum of the function $F(\xi)$ which differs from zero only in the interval $-1 \leqslant \xi \leqslant 1$. In general, $F(\xi)$ is a complex function (or a real function taking on both positive and negative values). As a result, $f(\eta)$ cannot serve as an autocorrelation function. However, the magnitude, $|F(\xi)|$, is a positive function with a bounded length, and the Fourier transform of this function results in a new function $R(\eta)$, which differs from $f(\eta)$. This function can be taken as the autocorrelation function, since it has a positive and bounded spectrum.

It is important that the new function $R(\eta)$ have a shorter duration than the initial function $f(\eta)$. Consequently, by using this method to synthesize from a non-realizable autocorrelation function $f(\eta)$ (belonging to the W_1 class), it is always possible to find another function $R(\eta)$ which is realizable and "better" than the given function. The proof of this statement follows from the definition of duration that was used in §4 and §17. The energies of the functions $f(\eta)$ and $R(\eta)$ with spectra $F(\xi)$ and $|F(\xi)|$, respectively, are identical and equal:

$$E = \frac{1}{2\pi} \int_{-1}^{1} |F(\xi)|^2 \, d\xi .$$

In addition, we also have

$$|f(\eta)| = \left| \frac{1}{2\pi} \int_{-1}^{1} F(\xi) \, e^{j\eta\xi} \, d\xi \right| \leqslant \frac{1}{2\pi} \int_{-1}^{1} |F(\xi)| \, d\xi = R_{\max} .$$

Therefore,

$$T_f = \frac{E}{f_{\max}^2} \geqslant T_R = \frac{E}{R_{\max}^2} , \tag{24–13}$$

Q.E.D.

In §27 and §28, the most important (from the viewpoint of practical application) branch of synthesis theory will be considered, *i.e.*, the determination of the optimal correlation function having a minimum duration, as well as a few additional constraints. The proof given for (24–13) shows that if the optimal function belongs to the W_1 class, it also belongs to the W_1^+ subclass. If this were not so, there would exist a realizable function of shorter duration which contradicts the optimality conditions.

In the following, the general topic of signal synthesis using the autocorrelation function is divided into several independent sections. In §25 and §26, the synthesis of arbitrary autocorrelation functions is discussed, *i.e.*, the power spectrum of the signal is determined from a given $R(\eta)$. In §27, §28, and §30, the optimal, realizable autocorrelation function having the minimum duration for a given spectrum width, as well as some additional constraints, is determined and analyzed. In §29, the important case of synthesizing autocorrelation functions having a finite duration is considered. In §31, we proceed to the actual signal synthesis and consider the design of FM pulse signals from the power spectra. §32 and §33 are devoted to the design of PM signals. Subsequent sections of this chapter contain discussions on signal synthesis based on the behavior of the ambiguity function over the entire t, Ω-plane.

In conclusion, it is noted that the technique for synthesizing the autocorrelation function considered above is similar to a problem encountered in the theory of linear antennas. Such analogies have been encountered repeatedly (§10, §14, and §17), and in view of their particular importance to synthesis theory, some of the significant relationships are presented here.

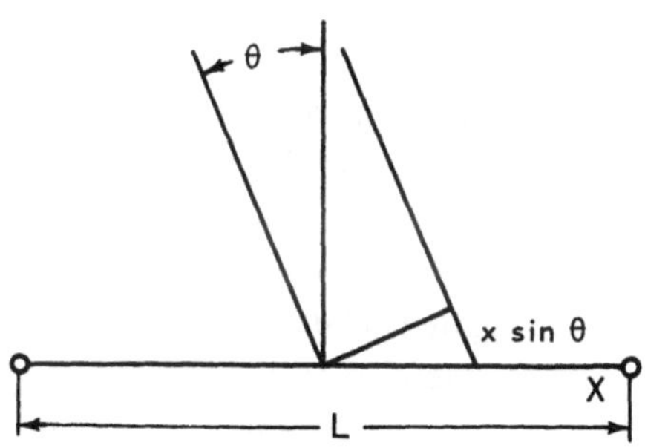

Fig. 27

Let us assume a linear wire antenna of length L has a current illumination $I(x)$ (see Fig. 27). Every increment of the antenna length radiates into space a field which is symmetrical with respect to the antenna axis. The total field is the resultant of the superposition of the elementary fields, and in the far field the individual components differ in phase only by the amount of their path length differences. Therefore, neglecting the unessential constant coefficient, the pattern of a linear antenna can be computed as follows:*

$$G(\theta) = \int_{-L/2}^{L/2} I(x)\, e^{j\frac{2\pi}{\lambda} x \sin\theta}\, dx .$$

Introducing the dimensionless coordinates

$$\xi = \frac{2x}{L} \quad \text{and} \quad \eta = \frac{\pi L}{\lambda} \sin\theta ,$$

* In the amplitude function of the integrand, the approximation $\cos\theta = 1$ is made.

we obtain, with additional normalization,

$$G(\eta) = \frac{1}{2\pi} \int_{-1}^{1} I(\xi)\, e^{j\xi\eta}\, d\xi. \tag{24–14}$$

This expression is obviously analogous to (24–7). Therefore, the signal synthesis problem is equivalent to the following problem: determine the current illumination, $I(x)$, for which the antenna pattern, $G(\eta)$ [defined by (24–14)], has the desired form. As in the case of signal synthesis, only the magnitude of the antenna pattern is usually given. The pattern is not given for all values of η but only in the interval,

$$-\pi\frac{L}{\lambda} \leqslant \eta \leqslant \pi\frac{L}{\lambda},$$

which corresponds to the real values of the angle θ. For long antennas, *i.e.*, $L/\lambda \gg 1$, (the synthesis theory is usually applied to such antennas) this restriction is not essential.

The synthesis of linear antennas has been studied for a comparatively long time and there is extensive literature on this topic (see, for example, a review by Fel'd & Bakhrakh [64]). Both the pencil-beam antenna with a narrow pattern and small sidelobes and the special shaped patterns (*e.g.*, the cosecant form) have been investigated. Naturally, many of the results obtained in the antenna study can be extended to the theory of signal synthesis.

In general, the current on the various antenna increments can have different phases, which results in a complex illumination $I(x)$. Frequently, complex illuminations are needed to obtain the desired patterns. Such pattern functions do not have analogs among the autocorrelation functions, since $F(\xi) = |S(\xi)|^2$ in the integrand of (24–7) is real and positive. However, there exists an important class of center-fed antennas for which the currents at every point on the antenna are in phase (usually designated as the zero phase). For such antennas, the function $I(x)$ is real and positive. Thus, the synthesis of autocorrelation functions is equivalent to the synthesis of antennas having a uniform phase distribution.

The table on p. 139 gives a comparison of the parameters encountered in signal synthesis with those encountered in antenna synthesis.

However, we note that in the theory of antennas, the integration is over finite limits due to their physical structure. The antenna dimensions are finite and, therefore, the antenna pattern must be obtained by integrating over finite limits.

No such limitations exist in the theory of signals. Here the requirement that the frequency band be finite is a mathematical idealization and results

Signal Parameters	Equivalent Parameters of a Linear Antenna Having Uniform Phase Distribution
Spectrum width Ω_m	Antenna length L
Dimensionless time units $\eta = \frac{\Omega_m t}{2}$	Dimensionless angular coordinate $\eta = \frac{\pi L}{\lambda} \sin\theta$
Autocorrelation function $R(\eta)$	Antenna pattern $G(\eta)$
Dimensionless frequency units $\xi = \frac{2\omega}{\Omega_m}$	Dimensionless antenna length units $\xi = \frac{2x}{L}$
Signal power spectrum $F(\xi) = \|S(\xi)\|^2$	Antenna current illumination $I(\xi)$

in some simplifications. The rejection of such an idealization does not contradict the physical formulation of the problem; to the contrary, this makes it possible to solve the synthesis problems from a broader class of physically realizable signals. However, from the practical viewpoint this difference is of no importance.

[§25] SYNTHESIS OF THE AUTOCORRELATION FUNCTION FROM ITS MAGNITUDE AND PHASE AT DISCRETE POINTS

Integral functions with a finite exponent (functions with a bounded spectrum) have one remarkable property. These functions are uniquely defined by their values over a series of discrete points. By assigning values to the function at various points, it is possible to define the function in the intervals between these points. This statement constitutes the *Kotel'nikov theorem* which can be applied to solving the synthesis problem. The proof of this theorem, which is well known (*e.g.*, see Khurgin & Yakovlev [71], p. 109), will not be covered in this book; however, its basic results are applied to the synthesis problem as follows.

The autocorrelation function $R(\eta)$ belonging to the W_1 class can be represented by the Kotel'nikov series:

$$R(\eta) = \sum_{k=-\infty}^{\infty} R(\pi k) \frac{\sin(\eta - \pi k)}{\eta - \pi k}. \tag{25–1}$$

This makes it possible to determine the values of the function for any η from the known discrete values at the points $\eta_k = \pi k$ $(k=0, \pm 1, \pm 2, \ldots)$.

The spectrum $F(\xi)$ of the function $R(\eta)$ differs from zero in the interval $-1 \leqslant \xi \leqslant 1$ and can be represented in this interval by the Fourier series

$$F(\xi) = \tfrac{1}{2} \sum_{k=-\infty}^{\infty} R(\pi k)\, \mathrm{e}^{-j\pi k \xi}, \tag{25–2}$$

whose coefficients are the values of the function $R(\eta)$ at the discrete points πk.

Evidently the Kotel'nikov theorem provides a simple method for the realization of autocorrelation functions. Both the series (25–1), which uniquely defines $R(\eta)$, and the Fourier series (25–2) are formed using the $R(\pi k)$ values. It can be shown that the series (25–2) yields a unique solution of the synthesis problem which coincides with the solution obtained using the Fourier integral method [*cf.*, (24–12)]. For the W_1 class of functions, such a solution is exact and unique. If the given function does not belong to the W_1 class, then it must be approximated beforehand by a function with a bounded spectrum. We shall now consider the method of Zelkin used in a similar antenna problem ([27], p. 88).*

Any continuous function can be approximated in a finite interval by a polynomial. The autocorrelation function is maximum at $\eta = 0$ and decreases rapidly as η increases. Therefore, it is always possible to select a region $|\eta| < c$ with a sufficiently large c so that the main beam of $R(\eta)$ is enclosed. Outside this region, the values of $R(\eta)$ are sufficiently small. Inside this chosen region the function $R(\eta)$ can be approximated by the polynomial

$$R(\eta) \approx P_n(\eta) = a_0 + a_1 \eta + \cdots + a_n \eta^n, \qquad -c < \eta < c.$$

We shall not dwell upon the methods for finding the polynomial coefficients, since these methods are thoroughly discussed in the literature (*e.g.*, see [14]). The important point is that by choosing the appropriate exponent n, it is possible to obtain the approximation with any given accuracy.

However, for large values of η, the polynomial $P_n(\eta)$ increases as $|\eta|^n$ and, consequently, differs substantially from the initial function $R(\eta)$ which decreases for large values of η. Therefore, $R(\eta)$ is approximated by the product

$$R(\eta) \approx U_m(\eta)\, P_n(\eta), \tag{25–3}$$

where the auxiliary function $U_m(\eta)$ has the following properties:

(1) for $|\eta| < c$, the function $U_m(\eta)$ differs negligibly from unity;

(2) for $|\eta| > c$, the function $U_m(\eta)$ decreases fast (at least faster than $|\eta|^{-n}$).

Thus, (25–3) yields a close approximation of the function $R(\eta)$. The

* In this section we do not consider the distinction between the W_1 and the W_1^+ classes. Keeping this in mind, it is possible to substitute $F(\xi)$ by $|F(\xi)|$ (see pp. 135 and 136).

function $U_m(\eta)$ can be taken as

$$U_m(\eta) = \frac{\sin \eta}{\eta \prod_{p=1}^{m} \left(1 - \frac{\eta^2}{\pi_2 p^2}\right)}. \tag{25–4}$$

Sin η can be represented by an infinite product:

$$\sin \eta = \eta \prod_{p=1}^{\infty} \left(1 - \frac{\eta^2}{\pi^2 p^2}\right).$$

Therefore,

$$U_m(\eta) = \prod_{p=m+1}^{\infty} \left(1 - \frac{\eta^2}{\pi^2 p^2}\right).$$

The function $U_m(\eta)$ has a maximum (which equals unity) when $\eta=0$ and decreases slowly around $\eta=0$ (for sufficiently large values of m). The first zero of $U_m(\eta)$ is at $\eta=\pi(m+1)$, and subsequently the zeros occur after every π interval. It can be shown that at $\eta_k=k\pi$, the function $U_m(\eta)$ has the following values:

$$U_m(k\pi) = \begin{cases} \frac{m!}{(m+k)!} \cdot \frac{m!}{(m-k)!} & \text{for} \quad |k| \leqslant m, \\ 0 & \text{for} \quad |k| > m. \end{cases} \tag{25–5}$$

In addition, it can be seen from (25–4) that for sufficiently large real values of η, we have the upper bound

$$|U_m(\eta)| \leqslant A_m |\eta|^{-(2m+1)},$$

where A_m is a constant.

Consequently, it is possible to chose for the function (25–4) a value of m so that both of the above conditions are satisfied, and the approximation (25–3) takes the form

$$R(\eta) \approx U_m(\eta)\, P_n(\eta) = \frac{\sin \eta P_n(\eta)}{\eta \prod_{p=1}^{m} \left(1 - \frac{\eta^2}{\pi^2 p^2}\right)}.$$

For such an approximation, the desired function becomes a member of the W_1 class. In fact, for large complex values of η, we have

$$\left| \frac{\sin \eta P_n(\eta)}{\eta \prod_{p=1}^{m} \left(1 - \frac{\eta^2}{\pi^2 p^2}\right)} \right| \leqslant C \frac{e^{|\eta|} |\eta|^n}{|\eta|^{2m+1}} = C \frac{e^{|\eta|}}{|\eta|^{2m-n+1}}$$

$$< C\, e^{(1+\varepsilon)\eta},$$

which corresponds to condition (24–11a). Consequently, to determine the sought-after spectrum of the autocorrelation function, we use (25–2) and (25–5); thus we have

$$F(\xi) \approx \frac{1}{2} \sum_{k=-m}^{m} \frac{m!}{(m-k)!} \cdot \frac{m!}{(m+k)!} P_n(\pi k) \, e^{-j\pi k \xi} . \tag{25–6}$$

Similarly, the autocorrelation function itself is defined by the series

$$R(\eta) \approx \sum_{k=-m}^{m} \frac{m!}{(m-k)!} \frac{m!}{(m+k)!} P_n(\pi k) \frac{\sin(\eta - \pi k)}{\eta - \pi k} . \tag{25–7}$$

The summation (25–6) defines the unknown power spectrum, $F(\xi)=|S(\xi)|^2$, of an autocorrelation function in both magnitude and phase. Some considerations concerning the application of this expression are discussed in the next section. Another method for solving a similar problem is given in §29.

[§26] SYNTHESIS OF THE AUTOCORRELATION FUNCTION FROM ITS MAGNITUDE

By using the approximation of Zelkin considered in the previous section, it is possible, under more restrictive conditions, to synthesize signals from the autocorrelation function when only its magnitude is known. Since the phase of $R(\eta)$ is of no practical significance, such a synthesis method is of particular interest.

The autocorrelation function has both real and imaginary parts, *i.e.*,

$$R(\eta) = R_1(\eta) + jR_2(\eta);$$

the reader is reminded that from condition (24–11b), $R_1(\eta)$ is even and $R_2(\eta)$ is odd. Therefore, approximation (25–3) is equivalent to the following system of equations:

$$\begin{aligned} R_1(\eta) &= U_m(\eta) \, A_{2p}(\eta), \\ R_2(\eta) &= U_m(\eta) \, B_{2q+1}(\eta), \end{aligned} \tag{26–1}$$

where $A_{2p}(\eta)$ and $B_{2q+1}(\eta)$ are polynomials of even and odd powers in η, respectively.

These polynomials can be written in the form

$$\begin{aligned} A_{2p}(\eta) &= M_1 \prod_{\alpha=1}^{p} (\eta^2 - \eta_\alpha^2), \\ B_{2q+1}(\eta) &= M_2 \eta \prod_{\beta=1}^{q} (\eta^2 - \eta_\beta^2), \end{aligned} \tag{26–2}$$

where η_α and η_β are the respective roots (real or complex).

Let us assume that only the magnitude of the autocorrelation function is known. From condition (24–11b), the magnitude $|R(\eta)|$ is an even function, and it can be approximated by using the even power polynomial $D_{2\nu}(\eta)$:

$$|R(\eta)| = U_m(\eta)\, D_{2\nu}(\eta),$$
$$D_{2\nu}(\eta) = M \prod_{k=1}^{\nu} (\eta^2 - \eta_k^2). \qquad (26\text{–}3)$$

Thus, polynomials $A_{2p}(\eta)$, $B_{2q+1}(\eta)$, and $D_{2\nu}(\eta)$ must be related as follows:

$$D_{2\nu}^2(\eta) = A_{2p}^2(\eta) + B_{2q+1}^2(\eta),$$

i.e.,

$$M^2 \prod_{k=1}^{\nu} (\eta^2 - \eta_k^2)^2 = M_1^2 \prod_{\alpha=1}^{p} (\eta^2 - \eta_\alpha^2)^2 + M_2^2 \eta^2 \prod_{\beta=1}^{q} (\eta^2 - \eta_\beta^2)^2. \qquad (26\text{–}4)$$

By making use of the expression

$$(a^2 - b^2)^2 = (a^2 + b^2)^2 - 4a^2 b^2,$$

we transform one of the factors on the left-hand side of (26–4) to the form

$$(\eta^2 - \eta_\mu^2)^2 = (\eta^2 + \eta_\mu^2)^2 - 4\eta_\mu^2 \eta^2.$$

By substituting this value into (26–4), it is readily seen that the polynomials $A_{2p}(\eta)$ and $B_{2q+1}(\eta)$ can be chosen as follows:

$$A_{2p}(\eta) = M(\eta^2 + \eta_\mu^2) \prod_{k=1}^{\nu}{}' (\eta^2 - \eta_k^2),$$
$$B_{2q+1}(\eta) = -\, 2M\eta_\mu \eta \prod_{k=1}^{\nu}{}' (\eta^2 - \eta_k^2). \qquad (26\text{–}5)$$

Here the $\prod'$ symbol denotes that the product is taken with the $k = \mu$ factor being excluded. Consequently, we have found one possible combination for separating the autocorrelation function magnitude into real and imaginary parts. Using (25–6), we obtain the corresponding form of the spectrum $F(\xi)$.

Performing similar transformations for other factors on the left-hand side of (26–4), other pairs of polynomials for $A_{2p}(\eta)$ and $B_{2q+1}(\eta)$ and other spectra are obtained; moreover, all the correlation functions have the form

$$|R(\eta)| = U_m(\eta)\, D_{2\nu}(\eta).$$

It is also possible to use expression

$$(a^2 - b^2)^2 (c^2 - d^2)^2 = [(ac - bd)^2 - (ad - bc)^2]^2,$$

and to transform simultaneously two, three, etc., factors of the product (26–3). In each case, new values of the polynomials $A_{2p}(\eta)$ and $B_{2q+1}(\eta)$ satisfying (26–4) are obtained, as well as new forms of the spectrum. By sorting all possible combinations in this manner, it is possible to find all the spectra that correspond to the chosen polynomial $D_{2\nu}(\eta)$. Thus, the synthesis of an autocorrelation function from a given magnitude does not have a unique solution; nevertheless [for a given $D_{2\nu}(\eta)$], there are a limited number of solutions among which it is possible to find the most acceptable for practical realization.

Thus, we arrive at the following procedure for the synthesis:

(1) the given magnitude of the autocorrelation function $|R(\eta)|$ is approximated in the interval $|\eta|<c$ by the polynomial $D_{2\nu}(\eta)$ of the 2ν power;

(2) the number m is chosen sufficiently large (the order of c/π) so that $(2m+1)>2\nu$, and the approximation is formed by using (25–3) and (25–4);

(3) by using (26–5) and similar expressions, the roots of the polynomial $D_{2\nu}(\eta)$ are determined (this may require cumbersome numerical calculations), and possible separations of $D_{2\nu}(\eta)$ into real and imaginary parts are found;

(4) for each of these separations, the power spectra $F(\xi)$ are determined by (25–6).

It is necessary to repeat the second step of the synthesis, *i.e.*, to determine the signal from the power spectrum. Moreover, if $F(\xi)$ has negative values, then it is necessary to substitute the magnitude $|F(\xi)|$ for $F(\xi)$ beforehand. As was shown in §24, the signal function found by this operation will have an autocorrelation function with a shorter duration than the given function. The second step of the synthesis can be carried out in the following manner: approximate the magnitude $|S(\xi)|$ by a polynomial, and by considering the possible separations (combinations of real and imaginary components) determine all realizable signals. A set of signals corresponds to each spectrum $|F(\xi)|$ found by the first step, but the total number of signals which are realized by the autocorrelation function (its magnitude) is limited.*

Thus, we have presented a comparatively complete synthesis method which allows us to find, to a certain degree of approximation, the total number of possible solutions. However, it is clear that this method is very cumbersome; also, the final signal form and the number of possible solutions depend substantially on the choice of the approximating polynomial for $|R(\eta)|$ and, subsequently, for $|S(\xi)|$. Such analytical properties as the number of real and complex roots of the polynomial, as well as their distribution over the complex plane, frequently affect the approximation accuracy insignificantly; however, it is these factors that define the signal form. It is not altogether clear to what extent and in what manner the accuracy of approximating $R(\eta)$ can be sacrified in order to obtain a signal of simpler form. This method is also inadequate due to the fact that after all the cumbersome calculations are made for determining $F(\xi)$, it frequently becomes necessary to replace this

* However, another method, more appropriate to the physical problem, is used in §31.

function by $|F(\xi)|$ and, by the same token, to reject the initial input function $R(\eta)$. It is clear that under these conditions it is impossible to estimate correctly the accuracy of the approximation.

Subsequently, an essentially different technique will be investigated. Instead of seeking a method for approximating an arbitrary autocorrelation function, we determine the optimal autocorrelation functions (in the W_1^+ class) which have the shortest duration for the various side-lobe levels. As expected, for each of the selected autocorrelation functions, there exists a single spectrum $F(\xi)$ with optimal properties and, consequently, at least the first portion of the synthesis problem can be solved uniquely. Subsequently, it is only necessary to zero-in on the optimal spectrum form having the desired accuracy. Thus, by rejecting the synthesis of arbitrary autocorrelation functions, it is possible to obtain a simpler solution which ensures an optimal, all-purpose form of $R(\eta)$. This, of course, does not exclude that in some particular cases special autocorrelation functions are needed which differ from the optimal functions. For these cases, one can use the above method.

There is a similar situation in antenna theory in which analogous methods are used to design antennas with special (*e.g.*, cosecant) directional patterns; in antenna theory, the optimal design achieves the narrowest beamwidth for a given side-lobe level. It is noted that, all things being equal, the optimal $R(\eta)$ having the shortest duration is always an even function. In fact, the autocorrelation function is maximum at $\eta=0$, *i.e.*,

$$|R_{\max}| = R(0).$$

Therefore, if

$$R(\eta) = R_1(\eta) + jR_2(\eta),$$

where R_1 and R_2 are even and odd functions, respectively, then $R_{\max} = R_1(0)$.

Consequently, of the two functions $R(\eta)$ and $R_1(\eta)$, the one with the lower energy also has the shorter duration. The addition of the imaginary component increases the energy:

$$E = \int_{-\infty}^{\infty} (R_1^2 + R_2^2)\, d\eta = E_1 + E_2 \geqslant E_1 ;$$

in this connection, the optimality condition is incompatible with the presence of the imaginary component.

[§27] OPTIMAL AUTOCORRELATION FUNCTIONS—LINEAR APPROXIMATION

Usually, when a signal is to be synthesized, very little is known about the sought-after autocorrelation function. As a rule, only the following two things are known: first, to improve the accuracy and resolution of the range measurement, the autocorrelation function must have as short a width as possible; second, to increase the discernment of the primary target from the nearby camouflaging targets, the function must have low side lobes. More detailed information on the form of $R(t)$ is not available. For the purposes of synthesis, one must select an $R(t)$, and here, naturally, there is substantial arbitrariness which significantly affects the quality of the approximation and the practical realization of the signal.

Under these conditions one must attempt to find an optimal, realizable autocorrelation function having the lowest side-lobe level for a given width of the principal beam and to determine the optimal signal spectrum which realizes this function. Ultimately, it is necessary only to approach this optimal, definitely-known signal form.

In stating the problem formally, it is necessary to define the optimality conditions more precisely, *i.e.*, to define accurately what is understood by the width of the autocorrelation function and the side-lobe level. Quite different and fairly arbitrary definitions are possible for these parameters. The width of the principal beam can be measured at various levels or determined for a rectangular pulse of equivalent energy (as is done in §4 and §17). The side-lobe level can be defined as the maximum level of $R(t)$ outside the main beam or, for that portion of signal energy outside a given duration. Different definitions of these quantities result in different optimal functions, $R(t)$, but it can be expected that such differences are not substantial. We shall illustrate this fact by considering two qualitatively different optimality criteria. One of these criteria is investigated in this section.

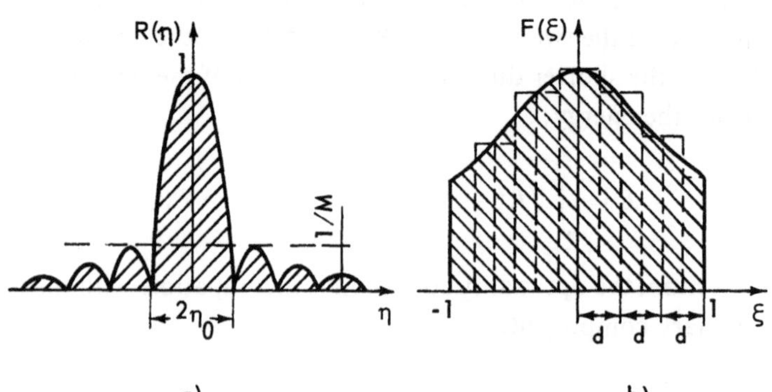

Fig. 28

Figure 28a illustrates a typical autocorrelation function $R(\eta)$ (in the following, the dimensionless coordinates $\eta = t\Omega_m/2$ and $\xi = 2\omega/\Omega_m$ are used again; Ω_m is the width of the signal spectrum) which is related to the power spectrum, $F(\xi)\equiv|S(\xi)|^2$, by

$$R(\eta) = A\int_{-1}^{1} F(\xi)\,e^{j\eta\xi}\,d\xi\,. \tag{27–1}$$

(Here the normalization multiplier A has been introduced to make the maximum value of the autocorrelation function, $R(0)$, equal unity.) The width of $R(\eta)$ is measured along the base of the main lobe, *i.e.*, between the first zeros of $R(\eta)$. The side-lobe level is characterized by M, which is the ratio of the amplitude of the main beam to the peak of the highest sidelobe ($M>1$). It is required to determine the autocorrelation function $R(\eta)$ and the corresponding function $F(\xi)$ which, for a given width η_0 has the minimum side-lobe level, $1/M$ (or conversely, for a given side-lobe level, has the minimum width). From the development presented in §26, it is assumed that $R(\eta)$ is an even function; thus, $F(\xi)$ is also an even function.

Let us divide $F(\xi)$ into $2n$ increments, each having a width $d=1/n$; also, let us assume that within each increment, $F(\xi)$ is approximately constant and equals $F_k(-n\leqslant k\leqslant n)$. Moreover, by virtue of the evenness,

$$F_{-k} = F_k\,.$$

Then, from (27–1) we obtain

$$R(\eta) = A\sum_{k=-n}^{n} F_k \int_{-\frac{d}{2}}^{\frac{d}{2}} e^{j\eta(\xi-kd)}\,d\xi$$

$$= 2A\,\frac{\sin\eta\,\frac{d}{2}}{\eta}\left(F_0 + 2\sum_{k=1}^{n} F_k\cos k\,d\eta\right). \tag{27–2}$$

Let us determine the normalization coefficient A. We have

$$R(0) = Ad\left(F_0 + 2\sum_{k=1}^{n} F_k\right).$$

If we impose the additional normalization condition on F_k,

$$F_0 + 2\sum_{k=1}^{n} F_k = \sum_{k=-n}^{n} F_k = 1\,, \tag{27–3}$$

and require that $R(0)$ be equal to unity, then we obtain

$$A = \frac{1}{d}.$$

As a result, (27–2) takes the form

$$R(\eta) = \frac{\sin \eta \frac{d}{2}}{\eta \frac{d}{2}} \left(F_0 + 2 \sum_{k=1}^{n} F_k \cos k\, d\eta \right). \tag{27–4}$$

In order to make the transition from the stepwise function $F(\xi)$ to a continuous function, we let $d \to 0$, $n \to \infty$, in such a manner that the product nd will remain unity. Moreover, the multiplier in (27–4) will also approach unity and, because of this, (27–4) becomes

$$R(\eta) = F_0 + 2 \sum_{k=1}^{n} F_k \cos k\, d\eta . \tag{27–5}$$

In this representation, the autocorrelation function $R(\eta)$ has the period $2\pi/d$ with respect to the variable η. By letting $d \to 0$, we increase the period without limit and $R(\eta)$ becomes a non-periodic function. This operation is analogous to the usual transition from a series to the Fourier integral.

The right-hand side of (27–5) is a polynomial to the nth degree in the variable

$$z = \cos \eta d ; \tag{27–6}$$

moreover, the half-period of $R(\eta)$ corresponds to the interval $-1 \leqslant z \leqslant 1$. Therefore, the problem of finding the optimal autocorrelation function formulated above reduces to determining an nth degree polynomial which equals unity for $z=1$ (*i.e.*, for $\eta=0$), and whose level outside the principal beam is minimum. It is well known that the Chebyshev polynomial given below has a similar characteristic:

$$T_n(\alpha z) = \cos(n \cos^{-1} \alpha z) = \cos(n \cos^{-1} x), \tag{27–7}$$

where the scale factor α must be determined separately. If $|x| > 1$, then by virtue of the function realizability, $T_n(x)$ must have the form (for real values of x)

$$T_n(x) = \cosh(n \cosh^{-1} x).$$

It is clear that in this region, $T_n(x)$ increases monotonically.

It is not difficult to see that all zeros of the polynomial $T_n(x)$ lie in the interval $-1 < x < 1$, whereas the factor $n \cos^{-1} x$ varies from 0 to $n\pi$ in this interval; $\cos(n \cos^{-1} x)$ becomes zero n times at the points

$$x_m = \cos \frac{2m-1}{2n} \pi ; \quad m = 1, 2, \ldots, n .$$

Consequently, the number of zeros equals the degree of the polynomial.

Between the zeros, at the points

$$x'_m = \cos\frac{m\pi}{n}; \quad m = 1, 2, \ldots, (n-1),$$

the Chebyshev polynomial takes on both maximum and minimum values; moreover, all maxima of $T_n(x)$ equal $+1$, and all the minima equal -1. Thus, the "side lobes" of $T_n(x)$ have the same magnitude.

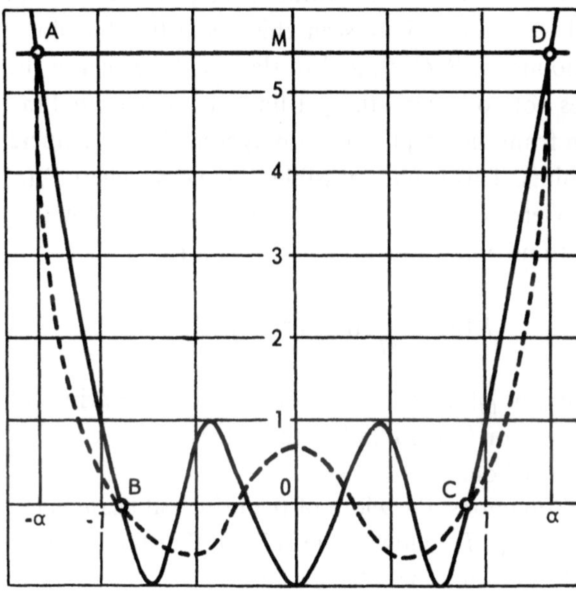

Fig. 29

A plot of $T_n(x)$ for the case of $n=6$ (solid line) is given in Figure 29. In order to make the transition from the Chebyshev polynomial to the optimal autocorrelation function, it suffices to change just the scale.

Let it be assumed that the principal maximum of $R(\eta)$ must exceed the side-lobe level by the factor M. We draw a horizontal line at the height M (Fig. 29). Since the side lobes of $T_n(x)$ are equal to unity, the intersections between this straight line with the plot of $T_n(x)$ define the scale factor α of (27–7). From the above discussion, this factor must satisfy (for $z=1$)

$$M = \cosh(n \cosh^{-1}\alpha),$$

i.e.,

$$\alpha = \cosh\left(\frac{1}{n}\cosh^{-1} M\right). \tag{27–8}$$

In addition, the scale along the vertical axis must also be changed by the factor M in order that the maximum of $R(\eta)$ remains equal to unity. Thus, using (27–6) and (27–7), we obtain

$$R(\eta) = \frac{1}{M} \cos[n \cos^{-1}(\alpha \cos d\eta)]. \tag{27–9}$$

It will be shown that it is this relationship which really defined the optimal (in the above sense) autocorrelation function. Assume that there is another function $R(\eta)$, corresponding to the nth degree polynomial $P_n(z)$ different from $T_n(\alpha z)$ so that, for the same width of the principal beam, it has a lower side-lobe level. As can readily be seen, this means that the plot of $P_n(z)$ passes through the points A, B, C, D, and in the side-lobe region (between points B and C) does not go beyond the ± 1 limits (shown dotted in Fig. 29). But, as can be seen from the graph, such a polynomial must coincide with $T_n(\alpha z)$ at the $n+2$ points. This is only possible in the case where $P_n(z) \equiv T_n(\alpha z)$.

It remains to let $n \to \infty$ and $d \to 0$ such that $nd = 1$. It follows from Appendix 4 that such a transition leads to the optimal form*

$$R(\eta) = \frac{1}{M} \cos \sqrt{\eta^2 - (\cosh^{-1} M)^2}. \tag{27–10}$$

Then, the values of $|\eta| < \cosh^{-1} M$ correspond to the principal beam region of the autocorrelation function, while the values of $|\eta| > \cosh^{-1} M$ correspond to the side-lobe region.

For optimal functions, the width of the principal beam is uniquely related to the side-lobe level. It follows from (27–10) that the relationships between the beamwidths and the side-lobe level are

$$\eta_0 = \sqrt{(\cosh^{-1} M)^2 + \frac{\pi^2}{4}} \approx \ln 2M,$$

$$\eta_{0.707} = \sqrt{(\cosh^{-1} M)^2 - \left(\cosh^{-1} \frac{M}{\sqrt{2}}\right)^2} \approx 0.84 \sqrt{\ln M};$$

the approximate relationships in the above are valid for large values of M.

The beamwidth—side-lobe level relationships are plotted in Figure 30.** From the above discussion, it can be seen that no autocorrelation function can have a principal beamwidth narrower than the side-lobe width at the

* The autocorrelation function (27–9) corresponds to the optimal pattern of an antenna array having n elements. Such antennas were first studied by Dolph and are frequently called the Dolph-Chebyshev arrays (Dolph [25], Khurgin and Yakovlev [71]). Eq. (27–10) corresponds to the optimal antenna pattern function of a continuous linear antenna. In antenna theory, this result was apparently first obtained by Van der Mass [46], but by a more complicated method. The method used in Appendix 4 was proposed by Sakolov [53].

** η_0 and $\eta_{0.707}$ are the one-half widths of the principal beam at the 0 and 0.707 levels, respectively.

same level (strictly speaking, this was only shown for the η_0 width). However, infinite energy is contained in the "non-attenuating" side lobes, so that the function which is optimum with respect to the side-lobe level and beamwidth is by no means optimum from the energy standpoint.

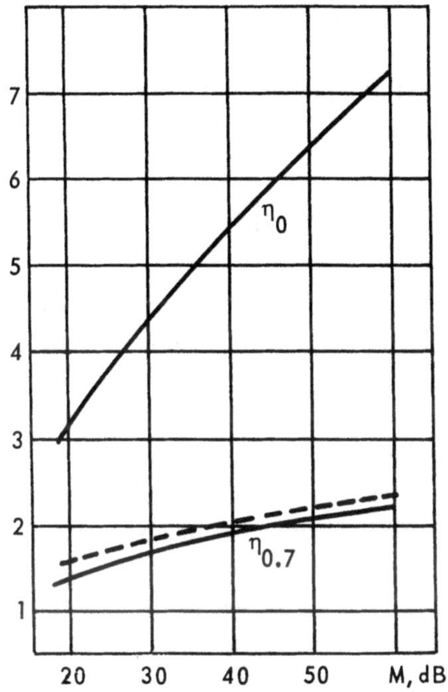

Fig. 30

Let us represent the autocorrelation function (27–10) in the form of the sum

$$R(\eta) = R_1(\eta) + R_2(\eta),$$

where

$$R_1(\eta) = \frac{\cos\sqrt{\eta^2 - (\cosh^{-1} M)^2} - \cos\eta}{M}. \tag{27–11}$$

Since $R_2(\eta)$ is the asymptotic expression of $R(\eta)$ as $\eta \to \infty$ [*cf.* (27.10)], the function $R_1(\eta)$ decreases for large values of η, and its side lobes attenuate comparatively rapidly. More accurately, it is not difficult to obtain the estimate

$$R_1(\eta) = \frac{(\cosh^{-1} M)^2}{2M}\,\frac{\sin\eta}{\eta} + O\left(\frac{1}{\eta^2}\right).$$

Therefore, $R_1(\eta)$ is integrable, and its Fourier transform $F_1(\xi)$ (the power spectrum of the signal) is bounded in magnitude and width [the latter follows from the fact that it can be readily verified that $R_1(\eta)$ belongs to the W_1

class]. Consequently, the optimal power spectrum $F(\xi)$ is

$$F(\xi) = \int_{-\infty}^{\infty} [R_1(\eta) + R_2(\eta)] \, e^{-j\xi\eta} \, d\eta$$

$$= F_1(\xi) + \pi \frac{\delta(\xi - 1) + \delta(\xi + 1)}{M}. \tag{27–12}$$

The δ-function sum in this equation is the Fourier transform of $R_2(\eta)$, which follows from the identity

$$\int_{-\infty}^{\infty} [\delta(\xi - 1) + \delta(\xi + 1)] \, e^{j\xi\eta} \, d\xi = e^{j\eta} + e^{-j\eta} = 2 \cos \eta .$$

Thus, the power spectrum of an optimal signal, $F(\xi)$, has infinite spikes (δ-functions) at the boundaries of the frequency band ($\xi = \pm 1$). This means that the optimal signal is not a pulse in the strict sense but also contains infinite sinusoidal components that are coherent with the pulse. Such a signal cannot be reproduced exactly. However, an abrupt increase of the spectrum $F(\xi)$, at the edges of the frequency band (and the addition of slowly decreasing components off the center frequency) yields, to a certain degree of approximation, the optimal autocorrelation function.

Of particular interest is the case where the power spectrum differs from the optimal spectrum by the complete absence of the spikes at the edge of the frequency band. In this case, the autocorrelation function is defined by (27–11) which, as a result of the normalization $R_1(0) = 1$, takes the form

$$R_1(\eta) = \frac{\cos \sqrt{\eta^2 - (\cosh^{-1} M)^2} - \cos \eta}{M - 1}. \tag{27–13}$$

Such autocorrelation functions are called *quasi-optimal* functions from the analogy with the corresponding antenna problem [53]. In Figure 31, there are plots of the optimal (solid line) and quasi-optimal (dotted line) autocorrelation functions constructed from (27–10) and (27–13). The dependence of the autocorrelation function principal beamwidth at the 0.707 level on M is also shown in Figure 30 (dotted line). As can be seen from the figures, the transition from optimal to the quasi-optimal function is accompanied by some widening of the width, and an increase of the close-in side lobes followed by a decrease of the remaining (far-out) side lobes. Quasi-optimal functions represent the limit of the worsening of the optimal function close to its center axis at the expense of lowering the "spikes" at the spectrum boundaries. However, quasi-optimal functions do not differ substantially from the optimal functions, and from the practical viewpoint, this difference can be frequently disregarded.

The quasi-optimal power spectrum $F_1(\xi)$ can be determined by calculating the Fourier transform of function (27–13):

$$F_1(\xi) = k \int_0^\infty [\cos\sqrt{\eta^2 - \beta^2} - \cos\eta] \cos\eta\xi \, d\eta \,, \qquad (27\text{–}14)$$

where $\beta = \cosh^{-1} M$ and k is the normalization factor defined in such a manner that $F(0) = 1$.

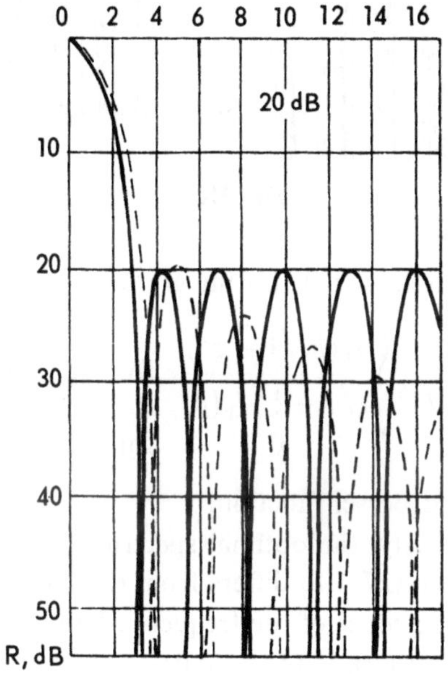

Fig. 31a

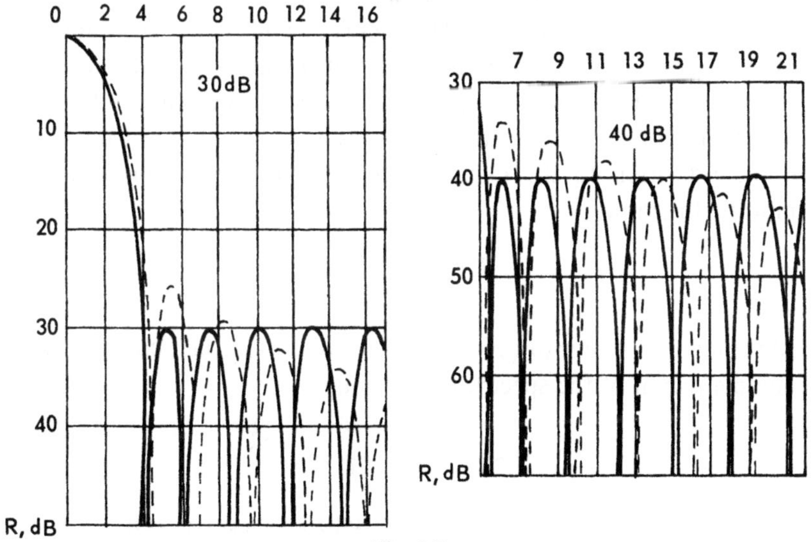

Fig. 31b

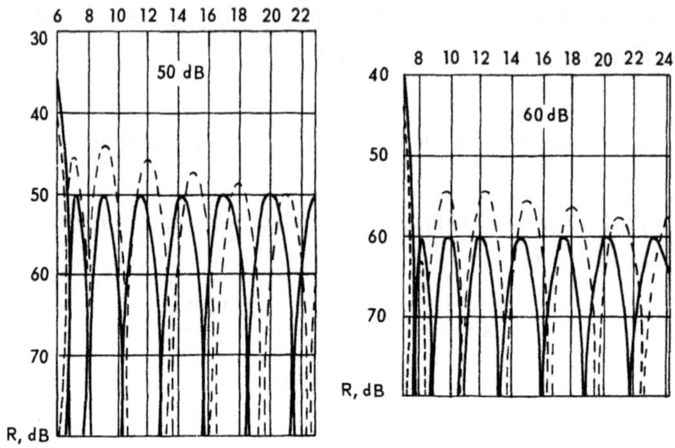

Fig. 31c

It is shown in Appendix 4 that this integration yields*

$$F_1(\xi) = \begin{cases} \dfrac{I_1(\beta\sqrt{1-\xi^2})}{\sqrt{1-\xi^2}\, I_1(\beta)} & \text{for} \quad |\xi| \leqslant 1, \\ 0 & \text{for} \quad |\xi| > 1, \end{cases} \tag{27–15}$$

where I_1 is the modified Bessel function.

As was noted previously, the optimal spectrum $F(\xi)$, which realizes the autocorrelation function (27–10), differs from (27–14) only by the presence of the δ-functions at the edges of the frequency band.

The plots of power spectra $F_1(\xi)$, calculated from (27–15), are shown in Figure 32. As can be seen from these curves, the spectra of the quasi-optimal pulses have the Gaussian form. These curves could have been anticipated because the spectra of Gaussian pulses are also Gaussian. Naturally, the use of Gaussian spectra with bounded widths leads to a decrease in the level of the lobes. If the given level of the side lobes is sufficiently low ($M \gg 1$, $\beta \gg 1$), then the transition to the Gaussian function follows formally from the asymptotic function

$$I_1(x) \sim \frac{e^x}{\sqrt{2\pi x}}.$$

Substituting this value into (27–15) and making $\xi \ll 1$ (*i.e.*, considering only the middle region of the spectrum), we obtain

$$F_1(\xi) \sim \frac{e^{-\beta(1-\sqrt{1-\xi^2})}}{(1-\xi^2)^{3/4}} \approx e^{-\beta\frac{\xi^2}{2}}.$$

* This expression is given by Van der Mass [46] and Minkovich [49] as the current distribution of a quasi-optimal linear antenna.

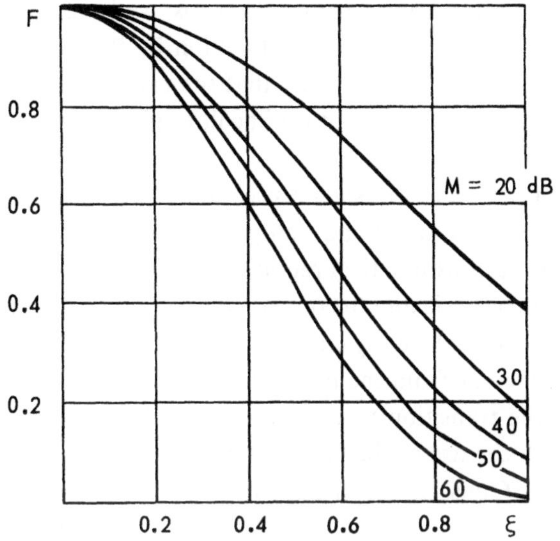

Fig. 32

If the side lobes are comparatively large ($\beta \lesssim 1$), then by using the approximate value

$$I_1(x) \approx \frac{x}{2},$$

we obtain a rectangular spectrum.

Thus, the frequently used signal with a rectangular spectrum (*e.g.*, rectangular pulses with a steep, linear FM) can be interpreted as a limiting case of optimal signals with a high side-lobe level. This means that for high side lobes it is not possible to obtain any substantial compression of the autocorrelation function (as compared to the case of the rectangular spectrum) at the expense of a special signal form. However, a transition to an optimal Gaussian spectrum makes it possible to decrease substantially the side-lobe level for a comparatively small increase in the principal beamwidth (*cf.* Figure 30). The quasi-optimal spectrum (27–15) can be physically realized, since $F_1(\xi)$ is a positive function bounded in width.

[§28] OPTIMAL AUTOCORRELATION FUNCTIONS—QUADRATIC APPROXIMATION

On the basis of the generalized Chebyshev polynomial, we have constructed an autocorrelation function with a minimum side-lobe level for a given width (with respect to the first zeros) of the principal beam. Such a definition of width has some objections since it is impossible to discriminate the zero position against a background of noise. In addition, the optimal signals satisfying this criterion are not realizable. Thus, we have gone to the quasi-optimal signal which eliminates the δ-function in $R(t)$. Although this was done after it was verified numerically that the exclusion of the δ-functions in $R(t)$ is beneficial (see Fig. 31), it is difficult to give a strict justification for such a transition.

The choice of a criterion in such a problem is always difficult. As a rule, it is not possible to formulate an "all-purpose" criterion which would be optimum for the various practical requirements. Therefore, it is appropriate to consider several criteria and to compare the solutions obtained from them. If these solutions are sufficiently close to one another, then the results can be considered satisfactory from the practical viewpoint.

In our case, the transition to the quasi-optimal autocorrelation function is associated with the use of the Gaussian power spectra. It can be expected that the use of the Gaussian spectrum is an "all-purpose" means of making it possible to narrow the principal beam of $R(t)$ and to decrease the side-lobe level for all "reasonable" definitions of these parameters. Therefore, it must be assumed that other criteria will also lead to similar results. We will confirm the above statement by considering the quadratic energy criterion of optimality.

The shorter the width of the autocorrelation function (or the compressed pulse), the wider its spectrum. However, it is possible to find an $R(t)$ with a given spectrum width which contains a major portion of its total energy in a given time interval. The criterion is a minimum principal beamwidth for a given spectrum; the beamwidth of the autocorrelation function $R(t)$ is defined as the time interval which contains the major portion of the energy. These autocorrelation functions (different for different energy ratios) will be called the *energy-optimal functions*.

The mathematical formulation of this method is as follows. It can be assumed, without loss of generality, that the total energy of $R(t)$ is equal to unity:

$$\int_{-\infty}^{\infty} |R(t)|^2\, dt = \frac{1}{2\pi} \int_{-\frac{\Omega_m}{2}}^{\frac{\Omega_m}{2}} F^2(\omega)\, d\omega = 1\,,$$

where $F(\omega) = |S(\omega)|^2$ is the power spectrum of the signal.

The portion of the energy contained in the interval $-T/2 < t < T/2$ is given by the integral

$$E_T = \int_{-\frac{T}{2}}^{\frac{T}{2}} |R(t)|^2\, dt\,. \tag{28–1}$$

Writing $R(t)$ in terms of its spectrum $F(\omega)$, and taking into account that $F(\omega)$ is a real function, we obtain

$$|R(t)|^2 = R(t)\,R^*(t) = \frac{1}{4\pi^2} \int_{-\frac{\Omega_m}{2}}^{\frac{\Omega_m}{2}} \int_{-\frac{\Omega_m}{2}}^{\frac{\Omega_m}{2}} F(\omega)\,F(\omega')\,e^{j(\omega-\omega')t}\,d\omega\,d\omega' .$$

Substituting the above into (28–1), we find

$$E_T = \frac{1}{4\pi^2} \int_{-\frac{T}{2}}^{\frac{T}{2}} dt \int_{-\frac{\Omega_m}{2}}^{\frac{\Omega_m}{2}} \int_{-\frac{\Omega_m}{2}}^{\frac{\Omega_m}{2}} F(\omega)\,F(\omega')\,e^{j(\omega-\omega')t}\,d\omega\,d\omega' ,$$

i.e.,

$$E_T = \frac{1}{2\pi} \int_{-\frac{\Omega_m}{2}}^{\frac{\Omega_m}{2}} \int_{-\frac{\Omega_m}{2}}^{\frac{\Omega_m}{2}} F(\omega)\,F(\omega') \frac{\sin(\omega-\omega')\frac{T}{2}}{\pi(\omega-\omega')}\,d\omega\,d\omega' . \qquad (28\text{–}2)$$

Thus, the problem of finding the optimal (with respect to energy) autocorrelation function reduces to determining the power spectrum $F(\omega)$ which results in a maximum of the integral (28–2) with the additional condition of

$$\frac{1}{2\pi} \int_{-\frac{\Omega_m}{2}}^{\frac{\Omega_m}{2}} F^2(\omega)\,d\omega = 1 . \qquad (28\text{–}3)$$

The parameters T and Ω_m are fixed quantities.*

From the calculus of variations, there exists a constant λ (the Lagrange multiplier) for which the solution $F(\omega)$ of (28–2) [plus the additional condition (28–3)] coincides with the solution of another maxima problem,

* The equivalent problem of a signal with a limited duration, having most of its energy confined to a given frequency band was solved by Gurevich [20, 21]. Other results based on the use of spheroidal functions (see below) were obtained by Sleipian, Landau, and Pollack [41], Vol. I, and also by Landgrebe and Cooper [39]. The forthcoming development is based mainly on the above works.

namely, the solution of the maximum of the function

$$J(F)=\int_{-\frac{\Omega_m}{2}}^{\frac{\Omega_m}{2}}\int_{-\frac{\Omega_w}{2}}^{\frac{\Omega_w}{2}} F(\omega)\,F(\omega')\,\frac{\sin(\omega-\omega')\,\frac{T}{2}}{\pi(\omega-\omega')}\,d\omega\,d\omega'$$

$$-\lambda\int_{-\frac{\Omega_m}{2}}^{\frac{\Omega_m}{2}} F^2(\omega)\,d\omega=\max.$$

Calculating the functional variation in the ordinary manner, it can be readily shown (*e.g.*, see Courant & Hilbert [36], p. 112) that the solution of this problem coincides with the solution of the following homogeneous integral equation:

$$\int_{-\frac{\Omega_m}{2}}^{\frac{\Omega_m}{2}} F(\omega')\,\varrho_T(\omega-\omega')\,d\omega'=\lambda F(\omega), \tag{28–4}$$

where

$$\varrho_T(\omega-\omega')=\frac{\sin(\omega-\omega')\,\frac{T}{2}}{\pi(\omega-\omega')}=\frac{1}{2\pi}\int_{-\frac{T}{2}}^{\frac{T}{2}} e^{j(\omega-\omega')t}\,dt. \tag{28–5}$$

Thus, instead of solving (28–2) with the restriction (28–3), it is possible to solve the homogeneous integral equation (28–4). This equation has solutions for only certain values of λ_i $(i=0, 1, 2, \dots)$ which are called the eigenvalues of the equation; the corresponding solutions, $F_i(\omega)$, are called the eigenfunctions. In our case, the eigenvalues, λ_i, have a simple physical meaning. Multiplying (28–4) by $(1/2\pi)\,F(\omega)$ and integrating with respect to ω, we obtain

$$\frac{1}{2\pi}\int_{-\frac{\Omega_m}{2}}^{\frac{\Omega_m}{2}}\int_{-\frac{\Omega_m}{2}}^{\frac{\Omega_m}{2}} F(\omega)\,F(\omega')\,\varrho_T(\omega-\omega')\,d\omega\,d\omega'=\lambda_i\frac{1}{2\pi}\int_{-\frac{\Omega_m}{2}}^{\frac{\Omega_m}{2}} F^2(\omega)\,d\omega.$$

According to condition (28–3), the total signal energy is equal to unity. The integral on the left-hand side of (28–2) is that portion of the energy, E_T, contained in the time interval $-T/2<t<T/2$; therefore,

$$E_T = \lambda_i . \tag{28–6}$$

It is necessary to maximize the energy and, consequently, to find the eigenfunction $F_0(\omega)$, which corresponds to the maximum eigenvalue ($\lambda_{\max}=\lambda_0$).

The optimal (in the energy sense) autocorrelation function is the Fourier transform of $F_0(\omega)$. Considering the case where i is arbitrary, we have

$$R_i(t) = \frac{1}{2\pi} \int\limits_{-\frac{\Omega_m}{2}}^{\frac{\Omega_m}{2}} F_i(\omega)\, e^{j\omega t}\, d\omega . \tag{28–7}$$

Using (28–4) and (28–5), we find

$$\begin{aligned} R_i(t) &= \frac{1}{\lambda_i} \frac{1}{2\pi} \int\limits_{-\frac{\Omega_m}{2}}^{\frac{\Omega_m}{2}} \int\limits_{-\frac{\Omega_m}{2}}^{\frac{\Omega_m}{2}} F_i(\omega')\, \varrho_T(\omega-\omega')\, e^{j\omega t}\, d\omega\, d\omega' \\ &= \frac{1}{2\pi\lambda_i} \int\limits_{-\frac{T}{2}}^{\frac{T}{2}} dt' \int\limits_{-\frac{\Omega_m}{2}}^{\frac{\Omega_m}{2}} \int\limits_{-\frac{\Omega_m}{2}}^{\frac{\Omega_m}{2}} F(\omega')\, e^{j\omega' t'}\, e^{j\omega(t-t')}\, d\omega\, d\omega' \\ &= \frac{1}{\lambda_i} \int\limits_{-\frac{T}{2}}^{\frac{T}{2}} R_i(t')\, \varrho_{\Omega_m}(t-t')\, dt' . \end{aligned}$$

Consequently, the $R_i(t)$ functions are eigenfunctions of the same integral equation,

$$\int\limits_{-\frac{T}{2}}^{\frac{T}{2}} R(t')\, \varrho_{\Omega_m}(t-t')\, dt' = \lambda R(t) , \tag{28–8}$$

as their spectra $F_i(\omega)$. Therefore, $R_i(t)$ and $F_i(\omega)$ have the same form. The difference between these two functions is due to the fact that the autocorrelation functions are expressed by (28–7) and (28–8) for all t, while the $F(\omega)$ spectra are different from zero in the frequency interval $|\omega| \leqslant \Omega_m/2$.

The eigenfunctions of (28–8) are called the *spheroidal functions.** It was shown by Slepian, Landau, and Pollak [41] that spheroidal functions have a number of useful properties and can be conveniently used for solving some complicated problems in the theory of signals. The most important of these properties are discussed in Appendix 5.

In this section we are only interested in the one particular spheroidal function which corresponds to the maximum eigenvalue, λ_0. In fact, this is the optimal (in the energy sense) autocorrelation function. The numerical data pertaining to this function given below are taken from [20, 39, 41].

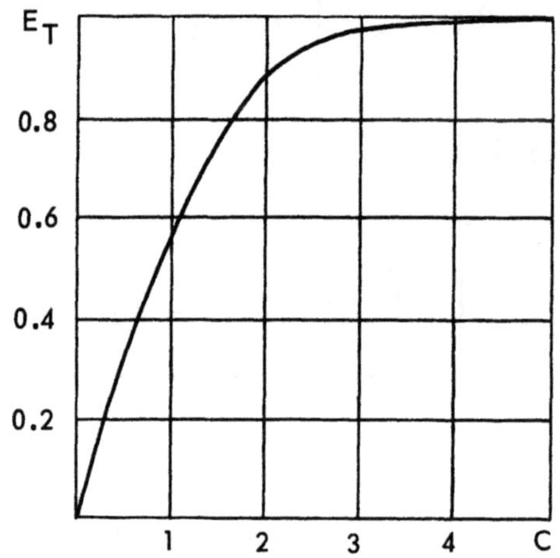

Fig. 33

Figure 33 shows the dependence of the partial energy E_T, contained in the $-T/2<t<T/2$ interval, on the parameter $c=T\Omega_m/2$. The optimal functions for some of the c values are given in Figure 34; Figure 35 has the plots of the corresponding $F(\xi)$ spectra. As can be seen from the plots, these functions have a clearly defined Gaussian shape for large values of c. It is shown in Appendix 5 that the asymptotic expressions

$$R(\eta) \sim e^{-\frac{\eta^2}{2c}} \quad \text{and} \quad F(\xi) \sim e^{-c\frac{\xi^2}{2}} \tag{28–9}$$

are valid for large values of c. The reader is reminded that

$$\eta = \frac{t\Omega_m}{2}, \quad \xi = \frac{2\omega}{\Omega_m}.$$

* These functions were first studied in connection with a boundary-value problem involving elongated spheroids (ellipsoids symmetrical with respect to the major axis) and hence the name.

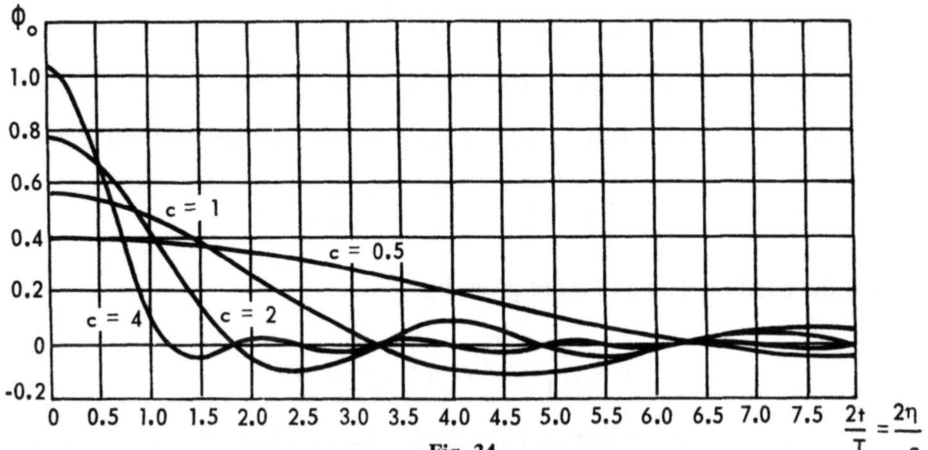

Fig. 34

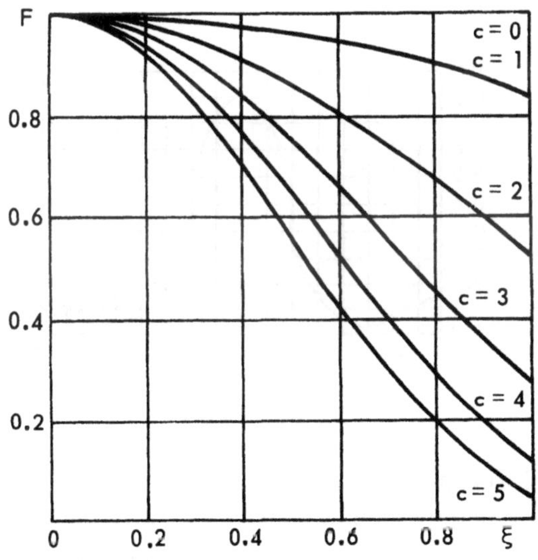

Fig. 35

Another result obtained with the aid of spheroidal functions will now be discussed. As was noted in §24, the requirement of a finite frequency band does not follow from the nature of the physical problem. It could be required that a certain portion of the energy, E_Ω, be contained in the frequency band of $|\omega| < \Omega_m/2$. What is the optimal autocorrelation function having the maximum possible energy in the time interval $|t| < T/2$ under this condition? Landau and Pollak ([41], pt. 2) showed that this condition is satisfied by the function

$$f = p\psi_0 + qD\psi_0,$$

where ψ_0 is the spheroidal function corresponding to the maximum eigenvalue λ_0 which, in turn, is defined by the equation

$$\cos^{-1}\sqrt{\lambda_0} = \cos^{-1}\sqrt{E_\Omega} + \cos^{-1}\sqrt{E_T}; \qquad (28\text{–}10)$$

$D\psi_0$ denotes that the function ψ_0 is bounded in the interval of $|t|<T/2$, *i.e.*,

$$D\psi_0 = \begin{cases} \psi_0 & \text{for} \quad |t| \leqslant T/2, \\ 0 & \text{for} \quad |t| > T/2, \end{cases}$$

and the coefficients p and q are given by

$$p = \sqrt{\frac{1 - E_T}{1 - \lambda_0}}, \quad q = \sqrt{\frac{E_\tau}{\lambda_0}} - p.$$

For the case of $E_T=0.977$ and $E_\Omega=0.96$, the function $f(t)$ is shown in Figure 36. The characteristic feature of such functions is the presence of the

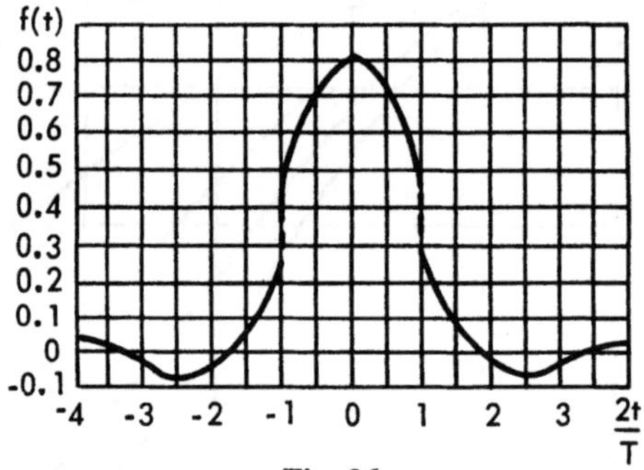

Fig. 36

steps at $|t|=T/2$. It can be readily shown that this is due to the fact that such autocorrelation functions cannot be realized, since the spectrum $F(\xi)$ alternates in sign (for large values of ξ); the condition for realizability, *i.e.*, $F(\xi)=|S(\xi)|^2 \geqslant 0$ is not satisfied.*

[§29] SYNTHESIS OF AUTOCORRELATION FUNCTIONS GIVEN OVER A BOUNDED TIME INTERVAL

The use of spheroidal functions makes it possible not only to investigate the energy optimal case but also to approximate arbitrary functions which do not belong to the W_1 class. It was previously noted that in practice the

* Equation (28–10), which characterizes optimal signals bounded (approximately) in frequency and time, was derived by the authors of [41] to formulate the uncertainty principle given in §13 (Fig. 22).

desired form of the autocorrelation function $R(\eta)$ can be finite only over a certain bounded time interval, *i.e.*, for $|\eta|<c$. Outside this interval, $R(\eta)$ must be as small as possible. Due to the above, it is expedient to formulate the synthesis technique as follows.

Let us assume that the function $R(\eta)$ is known in the interval $-c<\eta<c$, and, in general, that the function does not belong to the W_1 class; thus, $R(\eta)$ is not realizable. It is necessary to find the function $f(\eta)$ belonging to the W_1 class that gives the best approximation (in the sense of a least square fit) to $R(\eta)$ within the given interval, *i.e.*, the requirement is that the integral

$$\varepsilon = \int_{-c}^{c} |R(\eta) - f(\eta)|^2 \, d\eta \tag{29–1}$$

be at its minimum value. The function $f(\eta)$ contains some energy in this interval $(-c, c)$. However, it is also required that this energy be equal to

$$\int_{-\infty}^{\infty} |f(\eta)|^2 \, d\eta - \int_{-c}^{c} |f(\eta)|^2 \, d\eta = \delta . \tag{29–2}$$

If $R(\eta)$ is normalized with respect to energy, *i.e.*,

$$E_R = \int_{-c}^{c} |R(\eta)|^2 \, d\eta = 1 , \tag{29–3}$$

and $\delta \ll 1$, then condition (29–2) ensures a low level of the "side lobes" when $|\eta|>c$.

Introducing the Lagrange multiplier μ, we note that the sought-after function $f(\eta)$ results in the following sum being minimized:

$$J = \varepsilon + \mu\delta = \min . \tag{29–4}$$

We shall now make use of the spheroidal functions, $\psi_i(\eta)$, discussed in Appendix 5. The function $R(\eta)$ can be represented in the interval $(-c, c)$ by the series

$$R(\eta) = \sum_{i=0}^{\infty} a_i \psi_i(\eta); \; -c \leqslant \eta \leqslant c , \tag{29–5}$$

where

$$a_i = \frac{1}{\lambda_i} \int_{-c}^{c} R(\eta) \, \psi_i(\eta) \, d\eta .$$

Since the sought-after function $f(\eta)$ belongs to the W_1 class, the series

$$f(\eta) = \sum_{i=0}^{\infty} b_i \psi_i(\eta) \tag{29–6}$$

converges everywhere. Substituting these two series into (29–1) and (29–2) and taking into account the double orthogonality of spheroidal functions [(A5–5) and (A5–6) of Appendix 5], we find

$$\varepsilon = \sum_{i=0}^{\infty} \lambda_i |a_i - b_i|^2, \qquad \delta = \sum_{i=0}^{\infty} (1 - \lambda_i) |b_i|^2. \tag{29–7}$$

Therefore, the functional requirement (29–4) takes the form

$$J = \sum_{i=0}^{\infty} [\lambda_i (\alpha_i - b_i)(\alpha_i^* - b_i^*) + \mu (1 - \lambda_i) b_i b_i^*].$$

The parameters b_i and b_i^* which bring this function to a minimum yield the solution of the defined problem.* Differentiating J with respect to b_i^*, we find the minimum from the following equation:

$$\lambda_i (a_i - b_i) - \mu (1 - \lambda_i) b_i = 0;$$

hence

$$b_i = \frac{a_i \lambda_i}{\lambda_i + \mu (1 - \lambda_i)}. \tag{29–8}$$

The "side-lobe energy" (29–7) can be written as follows:

$$\delta = \sum_{i=0}^{\infty} \frac{|a_i^2| \lambda_i^2 (1 - \lambda_i)}{[\lambda_i + \mu (1 - \lambda_i)]^2}. \tag{29–9}$$

From the known a_i coefficients and the given δ, it is possible to find μ, and by using (29–8), the b_i coefficients in the expansion of the sought-after $f(\eta)$ function can be determined.

A simpler estimate of δ can be obtained from

$$\frac{\lambda_i + \mu (1 - \lambda_i)}{2} \geqslant \sqrt{\mu \lambda_i (1 - \lambda_i)}.$$

Therefore, using (29–9), we find

$$\delta \geqslant \frac{1}{4\mu} \sum_{i=0}^{\infty} |a_i|^2 \lambda_i = \frac{1}{4\mu}.$$

This last sum is the normalized energy E_R of (29–3).

* Since b_i contains both real and imaginary components, the conjugate quantities b_i and b_i^* must be considered as having independent arguments.

By using the (29–6) expansion of $f(\eta)$, it is easy to determine the signal power spectrum $F(\xi)$. From (A5–2), we have

$$F(\xi)=\begin{cases}\sum_{i=0}^{\infty} b_i j^{-i}\sqrt{\frac{2\pi c}{\lambda_i}}\,\psi_i(c\xi) & \text{for } |\xi|\leqslant 1,\\ 0 & \text{for } |\xi|>1.\end{cases} \tag{29–10}$$

If the real and imaginary parts of $R(\eta)$ are even and odd, respectively, then the spectrum $F(\xi)$ given by (29–10) is a real function. However, $F(\xi)$ may be negative for some values of $\xi<1$ and, consequently, $f(\eta)$ cannot be realized as the autocorrelation function. But if $R(\eta)$ has a positive spectrum in the frequency interval of $-1\leqslant\xi\leqslant 1$, then it can be expected that (29–10) is not negative in this region, since the approximations of $R(\eta)$ and its spectrum are closely related.*

[§30] COMPARISON OF OPTIMAL POWER SPECTRA

In §27 (where optimal autocorrelation functions based on the Dolph-Chebyshev criterion were considered) we have seen that for the asymptotic expansion of the low side-lobe level case, the optimal power spectrum has the form

$$F(\xi)\sim e^{-\beta\frac{\xi^2}{2}}, \tag{30–1}$$

where $\beta=\cosh^{-1}M\gg 1$. The spectrum of the function which is energy optimum has the same form. From (28–9),

$$F(\xi)\sim e^{-c\frac{\xi^2}{2}}, \tag{30–1′}$$

where the parameter c characterizes the product of the signal duration by the spectrum width.

The limiting cases for small β and for small values of c coincide; the autocorrelation functions of both cases have the following rectangular spectrum (compare Fig. 32 with 35):

$$F(\xi)=\begin{cases}1 & \text{for } |\xi|\leqslant 1,\\ 0 & \text{for } |\xi|>1.\end{cases} \tag{30–2}$$

* The synthesis method present here may also prove to be very useful in antenna theory, where the condition on the positiveness of the spectrum is not required. The antenna pattern $R(\eta)$ is always given only for positive angles which correspond to the condition of $|\eta|\leqslant\pi L/\lambda=c$. The values of $f(\eta)$ for $|\eta|>c$ make the antenna superdirective due to δ, the accumulated energy in the adjacent zone. Thus, (29–6) and (29–8) define the best approximation of the given antenna pattern having a fixed superdirectivity that is characterized by δ. Eq. (29–10) is the antenna current illumination. The author has no knowledge that spheroidal functions were previously used for the synthesis of antennas or signals.

A more comprehensive comparison of the optimal spectra is given in Figure 37, where $F(\xi)$ is plotted from (27–15), (A5–2), and (30–1) for the case of $c=\beta=4$. These curves are so close to each other that frequently their differences may be disregarded. It can be observed that the greatest difference

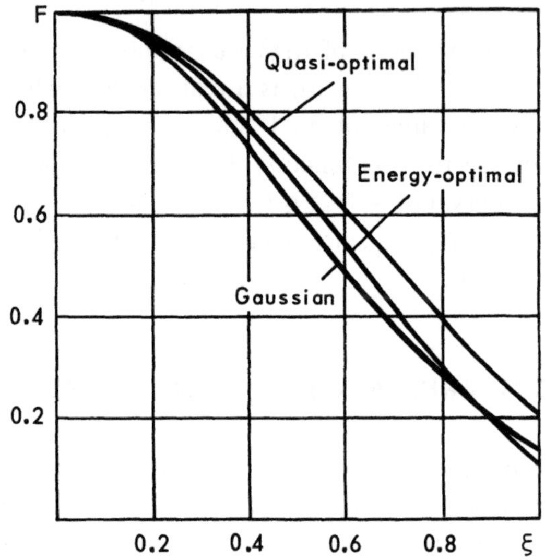

Fig. 37

occurs at the boundary of the frequency band, *i.e.*, at $\xi \approx 1$. This is as expected, since the step in magnitude at $\xi \approx 1$ defines the amplitude and the rate of remote side lobes decrease and, consequently, the energy contained in them. Therefore, for an energy-optimal autocorrelation function, the magnitude of the step must be relatively small.

It is noted that for $\beta=4$, the side-lobe level $M \approx 28$ dB. For $c=\beta=3$, the deviations between the optimal spectra increases insignificantly to about 5–10% (the asymptotic Gaussian curve differs more noticeably); for this case, $M \approx 20$ dB. Since for $M \approx 13$ dB, the optimal spectra once again coincide (become rectangular), the greatest difference can be assumed to be 5–10%. Thus, there is satisfactory agreement between the spectra obtained on the basis of the two criteria considered.* The spectra $F(\xi)$ of the optimal autocorrelation functions are Gaussian functions truncated at $\xi=1$. Different optimality criteria result in different but similar functions. It is natural to

* The corresponding autocorrelation functions are not given here, but we can readily convince ourselves that they also are in good agreement by comparing the curves in Figures 31 and 34. For example, the energy-optimal autocorrelation function for $c=4$ and the Dolph-Chebyshev function for $\beta=4$ have the same side-lobe level (approximately 30 dB).

expect that using other functions which are also different but similar to the optimum will lead to analogous results. For example, one can use the cosine-squared law (frequently used in antenna theory),

$$F(\xi) = \cos^2 \frac{\pi}{2} \xi \tag{30–3}$$

or even a smoother function of the type

$$F(\xi) = \cos^3 \frac{\pi}{2} \xi .$$

The generalized cosine-square law, which makes it possible to have a "pedestal" (*i.e.*, a non-zero level at the edges of the frequency band where $\xi = \pm 1$), is called the Hamming function,

$$F(\xi) = a + (1 - a) \cos \pi\xi , \tag{30–4}$$

which transforms into (30–3) when $a = \frac{1}{2}$ and into a rectangular spectrum when $a = 1$. The following table is a comparison of autocorrelation functions using the above spectrum for $F(\xi)$ both with the ideal (and non-realizable) spectrum based on the Dolph-Chebyshev criterion,

$$R(\eta) = \frac{1}{M} \cos \sqrt{\eta^2 - (\cosh^{-1} M)^2} ,$$

derived for the side-lobe level of $M = 40$ dB,* and the quasi-optimal function of (27–13) for the same value of M.

Type of function	Parameter a	Expansion coefficient of principal beam (at 3 dB level)	Level of highest side lobe dB	Decrease function of side lobes	Energy loss due to weighting process dB
Dolph-Chebyshev	–	1.00	40	No decrease	–
Rectangular	1	0.71	13.2	$1/t$	0.00
Hamming	0.58	1.00	34.2	$1/t$	1.01
	0.56	1.03	37.7	$1/t$	1.16
	0.54	1.06	42.8	$1/t$	1.34
	0.52	1.10	36.1	$1/t$	1.54
Cosine-squared	0.5	1.13	31.4	$1/t^3$	1.76
Cosine-cubed	–	1.34	39.1	$1/t^4$	2.38
Quasi-optimal	–	1.04	34.8	$1/t$	1.07

* These data are taken from the work of Temes [59].

The data given in the last column of the table need some additional explanation. It is relatively simple to form in the transmitter rectangular pulses with linear FM (assuming a sufficiently large compression ratio), and the processing leads to side lobes the order of 13 dB. To make the side lobes smaller, it is possible to "round the edges" of the receiver spectrum by using an IF amplifier with a Gaussian frequency characteristic. An FM pulse having a Gaussian spectrum (this pulse also has a Gaussian envelope) will be obtained at the receiver output; compressing this spectrum results in relatively small side lobes. Such a system of signal processing is, obviously, different from optimal and, consequently, leads to some energy loss and peak power reduction of a compressed pulse. This processing is usually called the weighted system, since the compressed pulse is represented by the Fourier integral

$$R(t) = \frac{1}{2\pi} \int w(\omega)\, F(\omega)\, e^{j\omega t}\, d\omega$$

with a weighting function (which corresponds to the frequency characteristic of the amplifier) $w(\omega) \leqslant 1$. It can be shown that the loss in the signal-to-noise ratio associated with such a processing is

$$\Delta\varrho = \frac{\left[\int_0^1 w(\xi)\, d\xi\right]^2}{\int_0^1 w^2(\xi)\, d\xi}. \tag{30–5}$$

This loss, expressed in dB, is given in the last column of the table. Computational difficulties only arise in the quasi-optimal characteristic when

$$w(\xi) = \frac{I_1\left(\beta\sqrt{1-\xi^2}\right)}{\sqrt{1-\xi^2}\, I_1(\beta)}.$$

Approximate calculations for this case are given in [53] from where the numerical data above were taken.*

As seen from the table, the "empirical" autocorrelation functions differ negligibly from the optimal functions. Moreover, with respect to most of the parameters given in the table, the quasi-optimal function at $M = 40$ dB is, obviously, no better than the Hamming function for values of a between 0.56 and 0.58. This, of course, is due to the fact that the "quasi-optimal" function was obtained in §27 under not very rigorous conditions and that the Dolph-Chebyshev functions serve as optimal.

* For the antenna problem considered in [53], the parameter $\Delta\varrho$ corresponds to the coefficient of antenna area utilization. This coefficient was determined graphically in [53].

[§31] SYNTHESIS OF FM SIGNALS FROM GIVEN POWER SPECTRA

Let us proceed to the second step in the synthesis, *i.e.*, to synthesize the signal proper. It is assumed that the power spectrum of the signal, $|S(\omega)|^2$, is determined from the autocorrelation function and that the requirement is to find the signal, $s(t)$, which satisfies the relationship

$$S(\omega) = \int_{-\infty}^{\infty} s(t)\, e^{-j\omega t}\, dt. \tag{31–1}$$

Since there is an arbitrariness in the phase of $S(\omega)$, the signal is not uniquely defined by such a method. This problem is not significantly different from that of synthesizing the autocorrelation function from a given magnitude. The signal synthesis problem is even simpler in that the spectrum $S(\omega)$ occupies a finite frequency band and, consequently, it is possible to use the Kotel'nikov series (see §26). However, it is possible to take advantage of the signal's phase arbitrariness to impose additional constraints which follow from engineering the desired signal structure. Here we shall consider the synthesis of a wide-band FM signal having a given power spectrum form.

By using the stationary phase method of Appendix 1, we have obtained a simple expression for the approximate spectra of wide-band FM pulses whose frequency changes monotonically. From this expression, the spectrum magnitude is [*cf.* (A1–8)]

$$|S(\omega)| = \sqrt{2\pi}\, \frac{a(t)}{\sqrt{|\omega_c'(t)|}}. \tag{31–2}$$

The reader is reminded that the stationary phase method yields only those frequency components of the spectrum which coincide with the instantaneous frequency of the pulse for its duration. Outside this region, the spectral density is considered to be zero.* Within this region, the spectral density function depends only on the behavior of the signal at the instant of time when the instantaneous frequency

$$\omega = \omega_c(t). \tag{31–3}$$

Eq. (31–2) shows that the spectrum magnitude depends on the amplitude, $a(t)$, and on the frequency rate of change, $\omega_c'(t)$, at this instant of time. Therefore, the spectrum must be matched to the required form in two ways: in both modulation amplitude and frequency.

* In reality, the spectral density outside this basic region decreases rapidly.

It is convenient to use the dimensionless variables in (31–2). As before, let us introduce the dimensionless frequency ξ, referenced to the mean frequency of the spectrum:

$$\xi = \frac{2(\omega - \omega_0)}{\Omega_{\max}}.$$

From the stationary phase method, the frequency band of the spectrum, $\Omega_{\max}$, equals the frequency deviation for the pulse duration, Ω_d; therefore, we also have

$$\xi = \frac{2(\omega - \omega_0)}{\Omega_d}; \quad -1 < \xi < 1.$$

Similarly, the instantaneous frequency expression is

$$\xi_c(t) = \frac{2(\omega_c(t) - \omega_0)}{\Omega_d}.$$

Contrary to ξ, the variable ξ_c depends on the time and varies over the interval $-1<\xi_c<1$ (the pulse duration). In addition, dimensionless time

$$\eta = \frac{\Omega_d}{2}(t - t_0),$$

referenced to time t_0 when the instantaneous frequency coincides with ω_0, is also introduced. If the signal pulse, $s(t)$, is symmetrical (an even function) having a duration T, then η is restricted to the interval

$$-\frac{\Omega_d T}{4} < \eta < \frac{\Omega_d T}{4} \quad \text{or} \quad -\frac{\pi}{2} m < \eta < \frac{\pi}{2} m,$$

where $m = \Omega_m T/2\pi$ is the compression ratio.

The instantaneous frequency boundaries $\eta = \pm\frac{1}{2}\pi m$ correspond to the time boundary values of $\xi_c(\eta) = \pm 1$. Using this notation, (31–2) can be written as

$$|S(\xi)|^2 = k\frac{a^2(\eta)}{\dfrac{d\xi_c}{d\eta}}. \tag{31–4}$$

The normalizing factor k introduced here will be determined subsequently. Eq. (31–4) is only valid for condition (31–3), which in the new notation has the form

$$\xi = \xi_c(\eta). \tag{31–5}$$

Considering also that $|S(\xi)|^2 = F(\xi)$, we find from (31–4) and (31–5) that

$$F(\xi_c)\, d\xi_c = k a^2(\eta)\, d\eta\,,$$

or

$$\int_0^{\xi_c} F(\xi)\, d\xi = k \int_0^{\eta} a^2(\eta)\, d\eta\,. \tag{31–6}$$

This equation establishes the relationship between the instantaneous signal frequency, ξ_c, and its envelope function, $a(\eta)$, with time η; it also makes it possible to synthesize FM signals having any given (arbitrary) power spectrum $F(\xi)$. In particular, if the optimal spectrum $F(\xi)$ based on the Dolph-Chebyshev criterion is given, then from (27–15) we obtain

$$\frac{1}{I_1(\beta)} \int_0^{\xi_c} \frac{I_1(\beta\sqrt{1-\xi^2})}{\sqrt{1-\xi^2}}\, d\xi = k \int_0^{\eta} a^2(\eta)\, d\eta\,. \tag{31–7}$$

Let us consider two special cases. Assume that it is required to synthesize an optimal FM signal having a rectangular envelope, *i.e.*,

$$a(\eta) = a_0\,, \qquad -\frac{\pi}{2} m < \eta < \frac{\pi}{2} m\,.$$

Then from (31–7), we obtain

$$k a_0^2 \eta = \frac{1}{I_1(\beta)} \int_0^{\xi_c} \frac{I_1(\beta\sqrt{1-\xi^2})}{\sqrt{1-\xi^2}}\, d\xi\,.$$

To find the factor k, let us resort to the energy relationship. By determining the signal energy in terms of its envelope function $a(\eta)$ and its power spectrum $F(\xi)$, we obtain

$$k \int_{-\frac{\pi}{2}m}^{\frac{\pi}{2}m} a^2(\eta)\, d\eta = \frac{1}{I_1(\beta)} \int_{-1}^{1} \frac{I_1(\beta\sqrt{1-\xi^2})}{\sqrt{1-\xi^2}}\, d\xi\,.$$

It can be shown (Watson [10]) that the last integral in this equation is equal to $(\cosh\beta - 1)/\beta$; hence, it follows that

$$2\frac{t}{T} = \frac{2\eta}{\pi m} = \frac{\beta}{\cosh\beta - 1} \int_0^{\xi_c} \frac{I_1(\beta\sqrt{1-\xi^2})}{\sqrt{1-\xi^2}}\, d\xi\,. \tag{31–8}$$

The reader is reminded that $\cosh\beta = M$ is the given side-lobe level of the autocorrelation function.

For this optimal (more accurately, quasi-optimal) case, the dependence of the instantaneous signal frequency ξ_c on time is shown in Figure 38. For a rectangular envelope, the Gaussian spectrum form is achieved at the expense of a non-uniform rate of change for the frequency. In the central

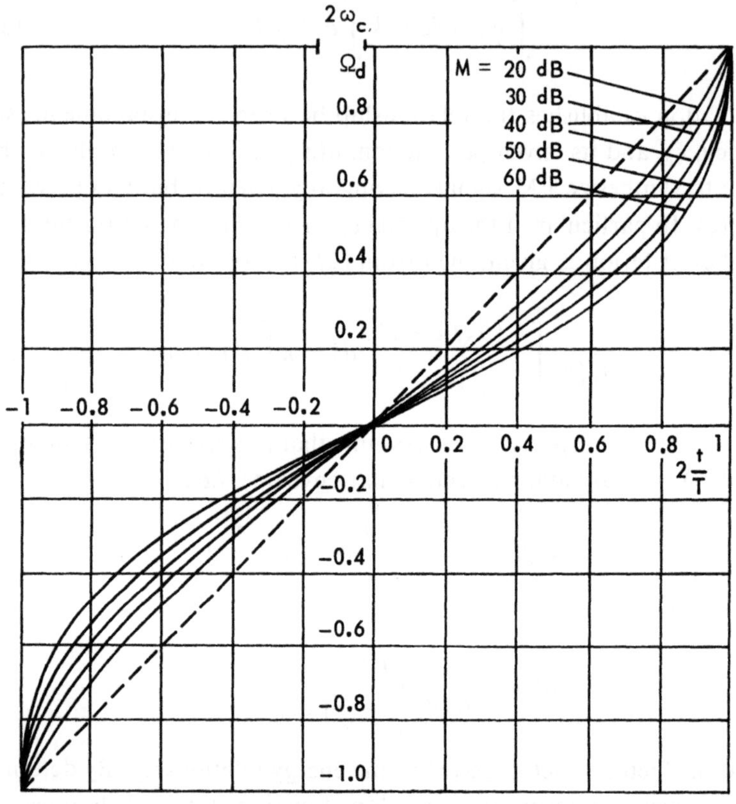

Fig. 38

portion of the pulse, the instantaneous frequency changes comparatively slowly, which leads to a relative increase in the energy concentrated about the center frequency. Near the edges of the pulse, the modulation rate increases and the spectral density decreases. The Gaussian spectrum form can be obtained not only by changing the frequency modulation rate but also by changing the pulse envelope to a special form.

For the second example, let us consider the case of having the frequency change linearly. Taking into account the scale of the dimensionless variables ξ_c and η, we must set

$$\xi_c = \frac{2}{\pi m}\eta = \frac{2t}{T}.$$

Differentiating (31–7), we find the sought-after amplitude modulation function

$$\alpha^2(\eta) = \frac{I_1\left(\beta\sqrt{1-\dfrac{4t^2}{T^2}}\right)}{\sqrt{1-\dfrac{4t^2}{T^2}}\, I_1(\beta)}, \qquad (31\text{–}9)$$

where, for convenience, the signal amplitude is normalized to unity. Thus, for linear FM, the required envelope form coincides with the spectrum form of the optimal signal. Figure 32 shows the square of this envelope.

We have found two FM signals, each of which approximately has a quasi-optimal autocorrelation function based on the Dolph-Chebyshev criterion. In the first case, this is achieved at the expense of requiring a special FM law for the rectangular envelope, and in the second case at the expense of a Gaussian envelope for linear FM. The synthesis problem for an arbitrary power spectrum $F(\xi)$ can be solved in an analogous manner. By assigning the signal envelope $a(\eta)$, it is possible to determine the FM law $\xi_c(\eta)$, and conversely by assigning $\xi_c(\eta)$, we can find the envelope $a(\eta)$.

Equation (31–6) or its equivalent (31–4) is the basic requirement for synthesizing FM signals, and for the special case of quasi-optimal signals, (31–7) is the requirement. These relationships were obtained by using the stationary phase method and thus are approximate. For a rigorous formulation, the problem requires the solution of

$$S(\omega) = \frac{1}{2\pi} \int_{-\frac{T}{2}}^{\frac{T}{2}} a(t)\, e^{j[\varphi(t)-\omega t]}\, dt,$$

in which the spectrum magnitude $|S(\omega)|^2 = F(\omega)$, and the envelope $a(t)$ are given. It is required to find the phase function $\varphi(t)$ or the instantaneous frequency $\omega_c(t) = d\varphi/dt$.* In fact, it is the approximate evaluation of this integral (using the stationary phase method) that leads to the result we have described. This approximation is suited for signals having a steep, monotonic FM law (for signals with large compression ratios), because it does not consider the "edge effect" associated with the pulse fronts. Of fundamental interest is the refinement of the solution to consider the "edge effects" (although it must be stipulated beforehand that a practical realization of a more accurate solution is doubtful). Such a correction method was developed by Tartakovskii and Tikhonova [57, 58] for application to antenna theory.

* Other versions are also possible here. For example, from a given FM law, find the amplitude envelope. It will be made clear in the following that the first case is of greater interest.

Let us consider the signal obtained by the stationary phase method [*i.e.*, satisfying (31–6)] as the initial approximation, *i.e.*,

$$s_0(t) = a_0(t)\,\mathrm{e}^{j\varphi_0(t)}$$

and let us determine its spectrum

$$S_0(\omega) = |S_0(\omega)|\,\mathrm{e}^{j\psi_0(\omega)} = \int_{-\frac{T}{2}}^{\frac{T}{2}} a_0(t)\,\mathrm{e}^{j[\varphi_0(t)-\omega t]}\,dt\,.$$

To evaluate this integral, numerical methods must be used. In practice, these computations are made on a computer. While the amplitudes of the frequency spectrum $F(\omega)$ and the given function $|S_0(\omega)|^2$ differ from each, they are still close. The signal envelope, $a_0(t)$, coincides with the given amplitude. In attempting to synthesize a signal having a given power spectrum, we take as the next approximation the spectrum,

$$S_1(\omega) = \{F(\omega)\}^{1/2}\,\mathrm{e}^{j\psi_0(\omega)}\,,$$

formed by adding the phase function of the initial approximation to the given amplitude spectrum. The first corrected signal approximation,

$$s_1(t) = a_1(t)\,\mathrm{e}^{j\varphi_1(t)} = \frac{1}{2\pi}\int_{-\frac{\Omega_m}{2}}^{\frac{\Omega_m}{2}} S_1(\omega)\,\mathrm{e}^{j\omega t}\,d\omega\,,$$

corresponds to the spectrum $S_1(\omega)$. Here, the integral has finite limits, since $F(\omega)$ differs from zero only in the interval $-\Omega_m/2 \leqslant \omega \leqslant \Omega_m/2$. Substituting $s_1(t)$ for the signal $s_0(t)$, the given power spectrum $F(\omega)$ will be realized exactly. However, the envelope $a(t)$ will not coincide with the given envelope $a_0(t)$. Therefore, we shall form the next (second corrected) approximation of the signal by taking the envelope $a_0(t)$ and assigning to it the phase of the first corrected approximation:

$$s_2(t) = a_0(t)\,\mathrm{e}^{j\varphi_1(t)}\,.$$

The spectrum corresponding to this signal,

$$S_2(\omega) = |S_2(\omega)|\,\mathrm{e}^{j\psi_2(\omega)} = \int_{-\frac{T}{2}}^{\frac{T}{2}} s_2(t)\,\mathrm{e}^{-j\omega t}\,dt\,,$$

differs (in absolute value) from the given $F(\omega)$. However, these differences are smaller than for the initial approximation [58]. Furthermore, it is possible to go on to the next approximation by adding the phase $\psi_2(\omega)$ to the given amplitude spectrum, etc.

Such an iteration process cannot claim a mathematical correctness and, what is more, this technique does not converge, since the spectrum with a bounded length cannot be realized by a signal of finite duration. However, it can be shown by calculation that even several steps yield a good approximation of the given $F_0(\omega)$ and $a_0(t)$ for practical application. Moreover, a correction for oscillations is introduced into the initial phase function $\varphi_0(t)$; this correction compensates for oscillations in the spectrum of the initial approximation associated with the fronts of the pulses. These oscillations cannot be accounted for within the framework of the stationary phase method.

[§32] PHASE-MODULATED SIGNALS

The generation of sophisticated signals is essentially limited to two cases due to practical considerations. Either FM pulses with large slopes (of frequency change) are used or phase-modulated (PM) signals, *i.e.*, pulses with multiple steps in the carrier phase.

Although it is feasible to realize the stepwise change in phase by any amount within the limits from 0 to 2π, it is mainly signals whose every step equals π that have practical application. There are two reasons for this: first, two identical pulses that differ in phase by π compensate one another and result in small side lobes of the autocorrelation function; second, such kinds of signals are comparatively simple to form (by switching the "polarity" at the appropriate instants of time). Some of the phenomena associated with the synthesis of such signals are considered in this and the following sections.

The PM signal consists of rectangular pulses having duration τ_0 (discretes), and can be written in the form

$$s(t) = \sum_{k=0}^{N-0} f(t - k\tau_0)\, e^{j(\omega_0 t + \pi\nu_k)} = \sum_{k=0}^{N-1} (-1)^{\nu_k} f(t - k\tau_0)\, e^{j\omega_0 t}, \tag{32–1}$$

where

$$f(t) = \begin{cases} 1 & \text{for} \quad 0 < t < \tau_0, \\ 0 & \text{for all other } t, \end{cases}$$

and the numbers ν_k $(k=0, 1, \ldots, N-1)$ are a sequence of zeros and ones which define the order of the phase alternation, *i.e.*, the signal code. It can be assumed that every signal consisting of N discretes is defined by the N-order binary number $\{v\}$. For example, the signal shown schematically in Figure 39a corresponds to the number 0001101. Thus, the synthesis of the PM signal is equivalent to choosing the binary number $\{v\}$.

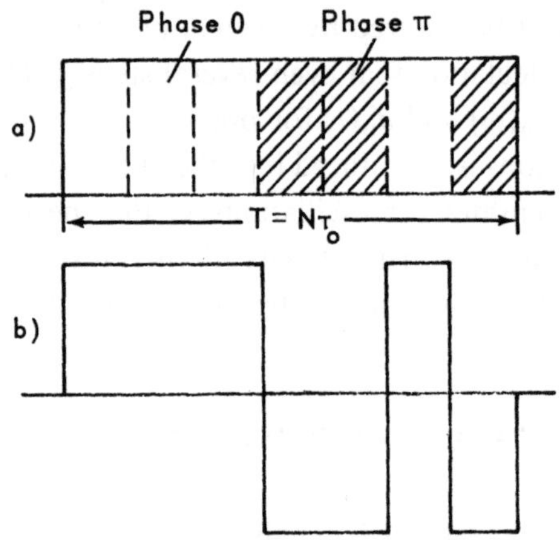

Fig. 39

Figure 39b illustrates the complex envelope of the PM signal of (32–1) with the $e^{j\omega_0 t}$ factor omitted. This envelope fully defines the signal spectrum (about the ω_0 frequency). Therefore, in the following, the signal envelope will be considered instead of the PM signal itself, *i.e.*, the video pulse as shown in Figure 39b. The complex envelope consisting of pulses having both polarities should not be confused with the magnitude $|s(t)|$ which, obviously, is a rectangular pulse having a total duration of $T=N\tau_0$ and does not depend on the signal code. The complex envelope [which will also be denoted by $s(t)$] can be expressed by

$$s(t) = \sum_{k=0}^{N-1} q_k f(t - k\tau_0), \tag{32–2}$$

where $q_k=(-1)^{v_k}=\pm 1$.

To illustrate this, the sequence of signs for the signal in Figure 39 is as follows: $+++--+-$. The synthesis can be considered choosing the sequence $\{q\}$.

In the following, we shall only consider the cases where all the discretes have the same duration τ_0, and the times of the steps, $t_1, t_2, \ldots$, are multiples of this duration. It may seem that such a condition unjustifiably narrows the class of signals, since the phase steps can be arranged arbitrarily for intervals which are not multiples of τ_0. In fact, this limitation is unessential since it is always possible to increase the number of discretes N by choosing τ_0 sufficiently small.*

The discrete nature of PM signals considerably complicates their synthesis. In the majority of synthesis methods, it is assumed that the sought-after coefficients must belong to a given class but take on a continuous sequence of values (forming a dense set). For PM signals, there are only two admissible values of q_k ($+1$ or -1), and the above-mentioned methods become unusable.

As an example, let us attempt to determine the PM signal $s(t)$ which approaches a certain continuous signal $s_1(t)$. Introducing the function

$$f_k(t) = f(t - k\tau_0) = \begin{cases} 1 & \text{for} \quad k\tau_0 < t < (k+1)\,\tau_0\,, \\ 0 & \text{for all other } t\,, \end{cases}$$

into (32–2), it can be readily seen that the $f_k(t)$ functions are orthogonal. In fact, the functions $f_k(t)$ and $f_i(t)$ do not overlap in time for $i \neq$ k. Therefore,

$$\int_{-\infty}^{\infty} f_k(t)\, f_i(t)\, dt = \begin{cases} 0 & \text{for} \quad i \neq k\,, \\ \tau_0 & \text{for} \quad i = k\,. \end{cases} \tag{32–3}$$

Consequently, (32–2) rewritten in the form

$$s(t) = \sum_{k=0}^{N-1} q_k f_k(t) \tag{32–4}$$

can be interpreted as the expansion of the signal $s(t)$ into the $\{f_k\}$ system. The coefficients q_k can be determined in such a manner that the difference $[s(t) - s_1(t)]$ will be as small as possible (in a certain sense). If the coefficients q_k could take on arbitrary complex values and if the number of terms in (32–4) are not rigorously restricted, then such a synthesis method would be quite satisfactory. However, in the case of PM signals, we must set $q_k = \pm 1$ for $0 \leqslant k \leqslant (N-1)$, and $q_k = 0$ for $0 > k \geqslant N$. Therefore, it is necessary to "round off" by setting the coefficients equal to $+1$ if $\mathrm{Re}(q_k) > 0$, and -1 if $\mathrm{Re}(q_k) < 0$; in addition, it is necessary to discard all the terms of the series that fall outside the interval $0 \leqslant k \leqslant (N-1)$. It is clear that for such an "approximation" it is difficult to obtain good results.

These difficulties practically exclude the synthesis of PM signals by ordinary methods. Therefore, in the recent literature, most of the attention has not been paid to the development of methods for synthesizing such signals from

* This argument is not valid if the times of the steps, t_1, t_2, ..., do not divide the duration into an integer number of increments. However, from the viewpoint of practical application, such a case is of no importance since by making arbitrarily small changes in t_1, t_2, ..., the duration can be divided into an integer number of increments.

arbitrary autocorrelation functions but rather to finding the actual signal itself which, if not optimum, is at least satisfactory. First, we shall consider two known classes of PM signals and present one generalization.

When the time $t=(k\tau_0+\tau)$, where $0<\tau<\tau_0$, the autocorrelation function of the PM signal can be represented by the form

$$R(k\tau_0+\tau)=\frac{1}{N\tau_0}\int_{-\infty}^{\infty} s(t')\, s(t'+k\tau_0+\tau)\, dt'$$
$$=\frac{1}{N\tau_0}\sum_{i=0}^{N-1} q_i \int_{-\infty}^{\infty} f_i(t')\, s(t'+k\tau_0+\tau)\, dt' .$$

Since $f_i(t')$ differs from zero only in the interval $i\tau_0 \leqslant t' \leqslant (i+1)\,\tau_0$, we find

$$R(k\tau_0+\tau)=\frac{1}{N\tau_0}\sum_{i=0}^{N-1} q_i \int_{i\tau_0}^{(i+1)\tau_0} s(t'+k\tau_0+\tau)\, dt'$$
$$=\frac{1}{N\tau_0}\sum_{i=0}^{N-1} q_i \int_{(i+k)\tau_0+\tau}^{(i+k+1)\tau_0+\tau} s(t)\, dt$$
$$=\frac{1}{N\tau_0}\sum_{i=0}^{N-1} q_i \Bigg[\int_{(i+k)\tau_0+\tau}^{(i+k+1)\tau_0} s(t)\, dt + \int_{(i+k+1)\tau_0}^{(i+k+1)\tau_0+\tau} s(t)\, dt\Bigg]$$
$$=\frac{1}{N\tau_0}\sum_{i=0}^{N-1} q_i [q_{i+k}(\tau_0-\tau)+q_{i+k+1}\tau]$$
$$=\frac{1}{N}\Bigg[\left(1-\frac{\tau}{\tau_0}\right)\sum_{i=0}^{N-k-1} q_i q_{i+k}+\frac{\tau}{\tau_0}\sum_{i=0}^{N-k-2} q_i q_{i+k+1}\Bigg]. \qquad (32\text{–}5)$$

It is easy to see that $R(t)$ consists of straight-line segments whose slope can change at the points $t=k\tau_0$. This broken line takes on maxima (in terms of absolute value) at the above points. Therefore, being interested in the magnitude of the lobes, only the discrete values need be considered:

$$R_k=R(k\tau_0)=\frac{1}{N}\sum_{i=0}^{N-k-1} q_i q_{i+k} . \qquad (32\text{–}6)$$

For the main lobe ($k=0$), $R(0)=1$ as expected, while for $k>0$, the sum (32–6) consists of $(N-k)$ terms, each of which equals either $+1$ or -1. The lobe minimum occurs when the sum is equal to zero. Obviously, the number of terms in (32–6) may be odd, and then the value of the sum can be either ± 1, *i.e.*,

$$|R_k| = \frac{1}{N}, \quad k \neq 0. \tag{32–7}$$

For a "best" sequence of $\{q\}$, the condition

$$\sum_{i=0}^{N-k-1} q_i q_{i+k} = \begin{cases} 0, & \text{if } N-k \text{ is even}, \\ \pm 1, & \text{if } N-k \text{ is odd}, \end{cases} \tag{32–8}$$

must be satisfied. It is noted here that $R(t)$ is an even function and, therefore, the negative values of k are not considered. One of the unknowns, q_i, can be chosen arbitrarily, for example by setting $q_0 = +1$. Then the above conditions form a system of $(N-1)$ equations with the following $(N-1)$ unknowns: $q_1, q_2, \ldots, q_{N-1}$. It is not possible to solve this system of equations by ordinary methods, since the unknowns can only assume two discrete values: $+1$ or -1.

The sequences $\{q\}$ satisfying condition (32–8) were found by trial and error for the values of $N = 2, 3, 4, 5, 7, 11, 13$ (see the table below).

n	k												
	0	1	2	3	4	5	6	7	8	9	10	11	12
2	+1	−1											
3	+1	+1	−1										
4	+1	+1	−1	+1									
5	+1	+1	+1	−1	+1								
7	+1	+1	+1	−1	−1	+1	−1						
11	+1	+1	+1	−1	−1	1	+1	−1	−1	+1	−1		
13	+1	+1	+1	+1	+1	−1	−1	+1	+1	−1	+1	−1	+1

These sequences (codes) are called the Barker codes [2]. Equation (32–8) is also satisfied for codes which are opposite to those given above, *i.e.*, for those having opposite signs, and for inverse codes which change the order of the signs (from right to left). All these codes are Barker codes.

The autocorrelation functions of Barker codes with $N = 7$ and $N = 13$ are shown in Figure 40. It is clear from these plots that PM signals with Barker code modulation have autocorrelation functions which, in a certain sense, are similar to the Dolph-Chebyshev optimal functions: both types of functions have the minimum possible side-lobe level (for a given class of signals). The maximum level of these side lobes is $1/N$.

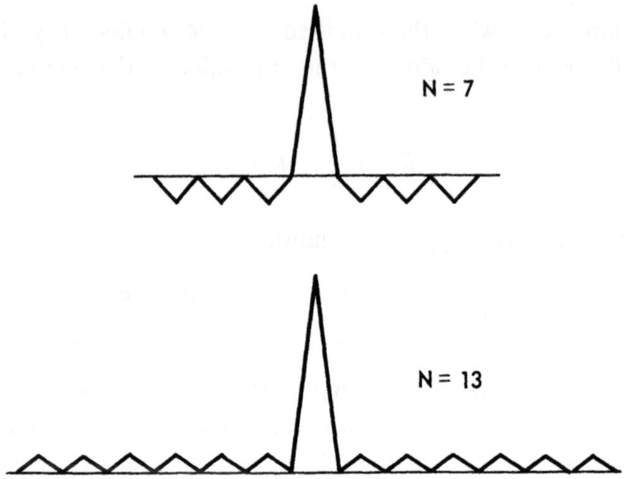

Fig. 40

We shall discuss another property of the Barker codes. It can be readily seen from (32–2) that the spectrum of PM signals can be written as

$$S(\omega) = \Phi_0(\omega) \sum_{k=0}^{N-1} q_k \, e^{-jk\omega\tau_0}. \qquad (32\text{–}9)$$

Here $\Phi_0(\omega)$ is the spectrum of one discrete increment and

$$\Phi_0(\omega) = \frac{2 \sin \omega \frac{\tau_0}{2}}{\omega} e^{-j\omega\frac{\tau_0}{2}};$$

the additional multiplier

$$Q(\omega) = \sum_{k=0}^{N-1} q_k \, e^{-jk\omega\tau_0}$$

is a function of the signal code. Obviously, $Q(\omega)$ is a periodic function with the period $\Omega = 2\pi/\tau_0$. For the power spectrum, we obtain

$$F(\omega) = |\Phi_0(\omega)|^2 \, |Q(\omega)|^2$$

moreover,

$$|Q(\omega)|^2 = Q(\omega)\, Q^*(\omega) = \left(\sum_{k=0}^{N-1} q_k \, e^{-jk\omega\tau_0}\right) \times \left(\sum_{i=0}^{N-1} q_i \, e^{ji\omega\tau_0}\right) = N + 2 \sum_{i=0}^{N-1} c_k \cos k\omega\tau_0 ,$$

where

$$c_k = \sum_{i=0}^{N-k-1} q_i q_{i+k}. \qquad (32\text{–}10)$$

The power spectrum $|Q(\omega)|^2$ contains a constant component N and sinusoidal terms whose amplitudes, c_k, depend on the signal code. If it is required that the "energy" contained in the sinusoidal terms be minimum, *i.e.*, that the integral

$$\int_{-\pi/\tau_0}^{\eta/\tau_0} [|Q(\omega)|^2 - N]^2 \, d\omega$$

have a minimum value, then the following condition must be satisfied:

$$\sum_{k=1}^{N-1} c_k^2 = \min .$$

It follows from (32–10) that for even values of $(N-k)$, the minimum value of c_k^2 is zero, while for odd values of $(N-k)$, c_k^2 cannot be smaller than unity. Thus, we arrive at the condition

$$c_k^2 = \begin{cases} 0, & \text{if} \quad N-k \quad \text{is even}, \\ 1, & \text{if} \quad N-k \quad \text{is odd}, \end{cases}$$

which is identical to (32–8).

It is clear from the above discussion that PM signals modulated by Barker codes have power spectra which deviate least (in the sense of a standard deviation measurement) from the spectrum of a single discrete. Accordingly, the autocorrelation functions of these PM signals approach the autocorrelation function of the discrete signal. The reader is reminded that the autocorrelation function of a rectangular signal is a triangular pulse similar to the principal spike in Figure 40.

Barker indicated that some codes produce a side-lobe level of $1/N$ only when $N \leqslant 13$. The use of sophisticated PM signals is more efficient for a large number of discretes, *i.e.*, when N is measured in hundreds or thousands. In this connection, numerous attempts have been made to find Barker codes for $N > 13$. However, the investigations of Ivanova, *et al.* [29], Turin and Storer [55, 61], and other authors have shown that no codes with a side-lobe level of $1/N$ exist for $N > 13$. Therefore, for large N, it becomes necessary to compose codes having a larger side-lobe level than $1/N$.

These kind of codes have been investigated by many authors (*e.g.*, see Siebert [28], Lerner [44], Hoffman [69], and the book by Lezin [43]). We shall now consider the class of codes which are called the M-sequences or zero-sequences of the maximum period.

As in the case of the Barker codes, the construction of the M-sequence is not solved by means of synthesis. Instead, according to the a priori algorithm, the sequence consisting of $+1$ and -1 is first constructed, and then its properties are investigated. In so doing, it becomes clear that the autocorrelation functions (and even the ambiguity functions) of the corresponding signals have one central peak and comparatively small side lobes. The algorithm for constructing the M-sequence reduces to the following [44]. An arbitrary set of numbers, $y_1, y_2, \ldots, y_n$, each of which is either zero or unity, and the initial sequence of numbers, $q_1, q_2, \ldots, q_{n-1}$, each of which is either $+1$ or -1, are given. The next term of the sequence q_n can be formed with the aid of the product

$$q_n = q_{n-1}^{y_1} q_{n-2}^{y_2} \cdots q_1^{y_{n-1}} .$$

Obviously, q_n equals either $+1$ or -1. The subsequent terms are determined by the recurrence relationship

$$q_{m+1} = q_m^{y_1} q_{m-1}^{y_2} \cdots q_{m-n+1}^{y_{n-1}} .$$

Thus, an infinite sequence of numbers $\{q\}$ is formed. It can be shown that after a certain number of terms, the sequence begins to repeat itself and the maximum repetition period is

$$N = 2^n - 1 .$$

Sequences with smaller periods can also be formed using methods from the theory of numbers; it can be shown that for any n there exists $\varphi(N)/n$ having the maximum repetition period. Here $\varphi(N)$ is the Euler function, *i.e.*, the number of prime integers in N but smaller than N. In fact, these are the sequences with the maximum period that are called the M-sequences; any sequence with such a period [containing $N=(2^n-1)$ signs] may be used as a code for the formation of a PM signal. It is clear that the finding of M-sequences contains procedures involving trial and error.

The basic properties of the M-sequences which are of importance in radar are as follows:

(1) It is possible to construct M-sequences whose period $N=(2^n-1)$ is greater than any given N_0. This means that by using an M-sequence, it is possible to form a PM signal with as large a number of discretes as desired.

(2) The autocorrelation function of a PM signal modulated by one period of the M-sequence has a side-lobe level the order of $1/\sqrt{N}$, *i.e.*,

$$R_k \lesssim \frac{1}{\sqrt{N}} . \tag{32–11}$$

(3) M-sequences have the property of being "random"; this property can be described as follows. Choose from the M-sequence period containing $N=(2^n-1)$ terms, all the intervals having n terms. It can then be seen that, first, there are no intervals among them that coincide, and second, there will

be among them many combinations of $+1$ and -1 consisting of n terms (except the "forbidden" combination containing only $+1$).

It is beyond the scope of this book to discuss the proofs of these and other properties of the M-sequences. For this purpose it would be necessary to use specialized mathematical methods not usually encountered in engineering and physics. Such a sophistication is not unexpected since the discrete nature of PM signals requires the use of the theory of numbers, which is the study of discrete quantities. In addition, methods for synthesizing PM signals having arbitrary properties are not known, and it is clear from the above discussion that we must use trial and error procedures when constructing M-sequences, consisting of $+1$ and -1, which are suitable for the formation of PM signals. It has been noted repeatedly that either by empirical methods, by combining several M-sequences, or, finally, by constructing other types of sequences, it is possible to find signals with a large number of discretes whose autocorrelation functions have a side-lobe level the order of $1/\sqrt{N}$. This suggests that there exists a substantially more general class of signals having a side-lobe level the order of $1/\sqrt{N}$ for large N. In the following, statistical methods will be used for obtaining such a class of signals, and the deterministic representations of the signal proper and the ambiguity function will be discarded. In particular, the side-lobe level will not be characterized by their maximum level, but by a statistical measure.

It will become apparent that the statistical approach of synthesizing PM signals makes it possible to obtain results which differ little from the "rigorous" ones, but this approach is substantially simpler and uses concepts which are familiar to engineers. By virtue of their "randomness", M-sequences are, apparently, special cases of this general class.

[§33] STATISTICAL SYNTHESIS OF PM SIGNALS

The following theorem of Wiener [11] is presented here in terminology more convenient for our purpose; this theorem serves as a starting point for the statistical formulation of the synthesis of PM signals: *Let the PM signal, $s(t)$, having unity amplitude consist of an infinite number of discrete increments, whose phases are random and independent of one another; moreover, the probability that the phase of the i-th increment is zero or π equals $1/2$. Then the autocorrelation function of such a signal, defined by*

$$R(t) = \lim_{N\to\infty} \frac{1}{2N\tau_0} \int_{-N\tau_0}^{N\tau_0} s(t')\, s(t' + t)\, dt',$$

of probability measure 0, *will not differ from the autocorrelation function of a single discrete*

$$R(t) = \begin{cases} 1 - \dfrac{|t|}{\tau_0} & \text{for} \quad |t| < \tau_0, \\ 0 & \text{for} \quad |t| > \tau_0. \end{cases}$$

This theorem shows that in the limiting case of infinite duration, a random sequence of phases leads to the desired result: an autocorrelation function having a width the order of τ_0 and with no side-lobes. The reservation that there may be other $R(t)$ of probability measure 0 is of no significance from the practical viewpoint. Such a reservation takes into account the theoretically possible cases, having negligible probability, in which, for a random choice of phases, regular signals can be formed, as for example, by a periodic sequence in which all even discretes have one phase and odd discretes have another phase.

It is natural to assume that the situation is similar for PM signals consisting of a large, but finite number of discretes. In other words, long random sequences must have satisfactory autocorrelation functions.

As we have seen in the previous section, the following equation can be used for estimating the maximum side-lobe rejection of the autocorrelation function

$$R_k = R(k\tau_0) = \frac{1}{N} \sum_{i=0}^{N-k-1} q_i q_{i+k}.$$

If the phases of the discretes are random and independent, then q_i and q_{i+k} are random variables taking on the values of ± 1 with the same probability; moreover, q_{i+k} does not depend on q_i (it is assumed that $k \neq 0$). Introducing the notation $N_k = (N-k)$, the last equation can be rewritten in the form

$$R_k = \frac{1}{N} \sum_{i=0}^{N_k - 1} x_i. \tag{33–1}$$

Here, $x_i = q_i q_{i+k}$ is obviously also a random variable taking on the values of $+1$ or -1 with the probability of 1/2. For different i, the x_i variables are independent. The problem consists of determining the probability that R_k will take on a certain given value. This is the simplest version of the random walk problem which can be described as follows (see Chandrasekhar [72]).

A point can move in steps along the x-axis. The length of each step, x_i, which equals unity, can occur with equal probability in both the positive and negative directions; moreover, the direction does not depend on the previous step. The problem is to determine after n steps the probability that the point will be in position X, where

$$X = \sum_{i=0}^{n-1} x_i ; \quad -n \leqslant X \leqslant n . \tag{33–2}$$

The solution of this problem is, of course, simple. In order for a point to be in the X position, it is necessary that it takes m_1 steps in the one direction and $m_2 = (m_1 - X)$ steps in the other direction. The total number of steps is $(m_1 + m_2) = n$ and, therefore, the probability of any such combination is

$$\left(\frac{1}{2}\right)^{m_1} \left(\frac{1}{2}\right)^{m_2} = \frac{1}{2^n} .$$

The number of combinations leading to X equals $C_n^{m_1}$; therefore,

$$W(X) = C_n^{m_1} \frac{1}{2^n} = \frac{1}{2^n} \frac{n!}{m!(n - m_1)!} .$$

Since it follows from the above that

$$m_1 = \frac{n + X}{2} ,$$

we have, finally,

$$W(X) = \frac{1}{2^n} \frac{n!}{\left(\frac{n + X}{2}\right)! \left(\frac{n - X}{2}\right)!} . \tag{33–3}$$

Note that X has the same parity as n, so that $(n+x)/2$ and $(n-x)/2$ are integers. As expected, this probability function has a binomial distribution. It can be shown [72] that the expected position of the point after n steps is

$$\bar{X} = 0 ,$$

and that the standard deviation is

$$\sigma_x = \sqrt{n} .$$

Of particular interest to the subsequent discussion is the case in which n is large and $X \ll n$. In this case, we obtain, with the use of Stirling's formula,

$$\ln n! \sim (n + \tfrac{1}{2}) \ln n - n + \tfrac{1}{2} \ln 2\pi$$

and from the expansion

$$\ln \left(1 \pm \frac{X}{n}\right) \approx \pm \frac{X}{n} - \frac{1}{2}\left(\frac{X}{n}\right)^2 ,$$

we obtain

$$W(X) = \sqrt{\frac{2}{\pi n}}\, e^{-\frac{X^2}{2n}}. \tag{33–4}$$

Hence, it follows directly that the standard deviation of the point's position equals $\sqrt{n}$. Thus, for a large number of steps, the probability distribution is Gaussian. This result also follows from the central limit theorem since the sum (33–2) consists of a large number of equivalent random terms.

It should be noted that the approximate expression (33–4) is in good agreement with the exact expression, even for a comparatively small number of steps. The following table (taken from [72]) shows a comparison of these expressions for the case of $n=10$.

X	W from Eq. (33–3)	W from Eq. (33–4)
0	0.24609	0.252
2	0.20508	0.207
4	0.11715	0.113
6	0.04374	0.042
8	0.00977	0.010
10	0.00098	0.002

The agreement between these two expressions is satisfactory not only for small values of X but also for $X \approx n$, which violates one of the requirements given in the previous section.

The solution of the random walk problem that has been presented can be readily extended to the problem of interest which is concerned with the side lobes of the autocorrelation function. By comparing (33–1) and (33–2), it becomes clear that the probability that the side-lobe level has some value R_k is given by

$$W(R_k) = \frac{1}{2^{N_k}} \frac{N_k!}{\left(\frac{N_k + NR_k}{2}\right)! \left(\frac{N_k - NR_k}{2}\right)!}$$

or

$$W(R_k) \approx \sqrt{\frac{2N^2}{\pi N_k}}\, e^{-\frac{N^2 R_k^2}{2N_k}}, \tag{33–5}$$

where $N_k = (N-k)$.

The root mean square (RMS) value of the k-th side lobe is

$$\sigma_k = \frac{1}{N}\sqrt{N_k} = \frac{1}{N}\sqrt{N-k} = \sigma_0 \sqrt{1 - \frac{k}{N}}, \tag{33–6}$$

where

$$\sigma_0 = \frac{1}{\sqrt{N}}.$$

It is clear from this equation that the RMS value of the side lobes decreases with distance from the main lobe. For near-in side lobes ($k \ll N$), we have

$$\sigma_k \approx \sigma_0 = \frac{1}{\sqrt{N}},$$

and for remote side lobes ($k \approx N-1$),

$$\sigma_k = \frac{1}{N}.$$

Thus, PM signals of finite duration with random phase alternations have autocorrelation functions whose "mean" side-lobe levels do not exceed $1/\sqrt{N}$. This fact constitutes the principal property of random sequences and makes it possible to synthesize PM signals on the basis of these sequences.

The M-sequence considered in the previous section also has a side-lobe level the order of $1/\sqrt{N}$. However, the M-sequence discussion was concerned with the highest level of the side lobes, whereas for the random sequence we have estimated only the RMS level of the side lobes. It can be expected that in the statistical synthesis, a large number of the autocorrelation function's side lobes will exceed the $1/\sqrt{N}$ value. It is difficult to determine which characterizes the signal quality more fully: the maximum or the mean value of the side lobes. The mean level of the side lobes is, apparently, a satisfactory measure since the detection of signals against a background of interfering reflections is a statistical problem. With respect to the mean level, the random and M-sequences are approximately equivalent. However, since autocorrelation functions are frequently characterized by the peak value of the side lobes, we shall direct our discussion on random sequences along these lines.

If the signal is formed by a random sampling of phases and this signal is subsequently used as a radar transmission, then the filtering employed at the receiver is matched to the very signal transmitted. The autocorrelation function of such a signal has a very definite form. From the previous discussion, the mean side-lobe level of this autocorrelation function does not exceed $\sigma_0 = 1/\sqrt{N}$. However, there is always some probability that one or several of the side lobes will substantially exceed the mean level. For a random sampling of phases, it is possible that the maximum side lobe level may prove to be too high, and the only remaining thing to do is to form (also in a random manner) another signal, hoping that this signal will have a lower side-lobe level.

It is clear that if the criterion of maximum side-lobe level is used in the statistical synthesis, then it is necessary to know the probability of the highest sidelobe not exceeding a given level for the specified realization of the random sequence (consisting of N discretes). If such a probability is sufficiently high (let us say, 0.7 to 0.8), then we can be sure that in the worst case a signal with the required properties will be formed after two or three trials. Consequently, in statistical and other methods of synthesizing PM signals, phase functions formed by random sampling are not to be excluded.

Since the side-lobe amplitudes follow the normal distribution (33–5), the probability that the k-th side lobe will not exceed a specified level R is given by the expression

$$P_k = \int_0^R W(R_k)\, dR_k = \sqrt{\frac{2}{\pi}} \frac{1}{\sigma_k} \int_0^R e^{-\frac{R_k^2}{2\sigma_k^2}}\, dR_k .$$

We are interested only in the cases where the probability P_k is sufficiently close to unity; under such a condition, the probability can be approximated by

$$P_k \approx 1 - \sqrt{\frac{2}{\pi}} \frac{\sigma_k}{R} e^{-\frac{R^2}{2\sigma_k^2}} . \tag{33–7}$$

The levels of the autocorrelation function's side lobes are statistically independent; therefore, the probability that all $(N-1)$ side lobes will not exceed R is expressed by the product

$$P = P_1 P_2 \ldots P_{N-1} .$$

Since every factor in the above is close to unity, we obtain from (33–7) and (33–6), with the higher order terms neglected,

$$P \approx 1 - \frac{\sigma_0}{R} \sqrt{\frac{2}{\pi}} \sum_{k=1}^{N-1} \sqrt{1 - \frac{k}{N}}\, e^{-\frac{R^2}{2\sigma_0^2 \left(1 - \frac{k}{N}\right)}} .$$

It is pointed out that the upper limit in this sum can be replaced by N, since the $k = N$ term is equal to zero. In addition, for a large number of terms, the summation can be replaced by an integral. Therefore, introducing the variable $x = k/N$, we obtain

$$P \approx 1 - \sqrt{\frac{2}{\pi}} \frac{\sigma_0 N}{R} \int_0^1 \sqrt{1 - x}\, e^{-\frac{R^2}{2\sigma_0^2 (1-x)}}\, dx$$

$$\approx 1 - 2 \sqrt{\frac{2}{\pi}} \frac{N \sigma_0^3}{R^3} e^{-\frac{R^2}{2\sigma_0^2}} . \tag{33–8}$$

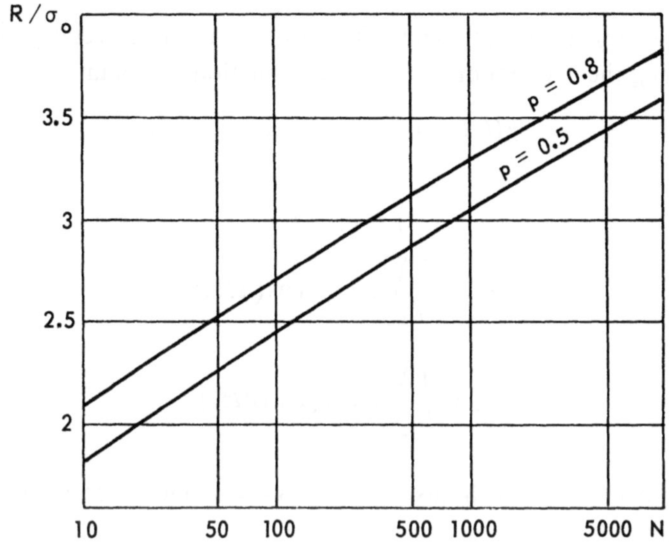

Fig. 41

Figure 41 shows the dependence of R/σ_0 on the number of discretes N computed from (33–8) for the probabilities of $P=0.5$ and $P=0.8$. It can be seen from this graph that the expected level of maximum rejection increases with respect to the mean level of $\sigma_0 = 1/\sqrt{N}$ as N increases. This increase is not too substantial, and in the most important region, the ratio R/σ_0 is from 2 to 4.

Let us also evaluate the *ambiguity function of a random signal over the entire t, Ω-plane*. The PM signal is

$$s(t) = \sum_{i=-\infty}^{\infty} q_i f(t - i\tau_0),$$

where the q_i coefficients are different from zero (equal ± 1) only in the interval $0 \leqslant i \leqslant (N-1)$.

Then, the ambiguity function is

$$\chi(t, \Omega) = \frac{1}{N\tau_0} \int_{-\infty}^{\infty} s^*(t')\, s(t' + t)\, e^{j\Omega t'}\, dt'$$

$$= \frac{1}{N\tau_0} \sum_i \sum_j q_i q_j \int_{-\infty}^{\infty} f^*(t' - i\tau_0)\, f(t' - j\tau_0 + t)\, e^{j\Omega t'}\, dt'.$$

After some simple transformations, we obtain

$$\chi(t, \Omega) = \frac{1}{N} \sum_i \sum_k q_i q_{i+k}\, e^{j\Omega i\tau_0}\, \chi_0(t - k\tau_0, \Omega),$$

where $\chi_0(t, \Omega)$ is the ambiguity function of a single rectangular pulse.

Changing the summation order in this equation, we obtain

$$\chi(t, \Omega) = \sum_k c_k(\Omega)\, \chi_0(t - k\tau_0 \Omega), \tag{33–9}$$

where

$$c_k = a_k + jb_k;$$

$$a_k = \frac{1}{N} \sum_i q_i q_{i+k} \cos(i\Omega\tau_0);$$

$$b_k = \frac{1}{N} \sum_i q_i q_{i+k} \sin(i\Omega\tau_0). \tag{33–10}$$

The sums over i have the limits of $0 \leqslant i \leqslant (N-k-1)$ when $k>0$, and $|k| \leqslant i \leqslant (N-1)$ when $k<0$; the variables c_k differ from zero when $-(N-1) \leqslant k \leqslant (N-1)$.

It can be readily perceived from these equations that the ambiguity function in the t, Ω-plane has the nature of hilly terrain. The function $\chi_0(t, \Omega)$ corresponding to one discrete, *i.e.*, to a rectangular pulse of duration τ_0 differs from zero only in the interval $-\tau_0 < t < \tau_0$. Therefore, in the sum (33–9), which consists of the functions $\chi_0(t, \Omega)$ shifted in time, not more than two terms overlap. In addition, each term is "modified" by a random multiplier $c_k(\Omega)$ along the frequency axis. Since the maxima of the shifted functions $\chi_0(t - k\tau_0, \Omega)$ fall on the zeros of the adjacent functions, the amplitudes of the "hills" are approximately defined by the magnitude

$$|c_k| = \sqrt{a_k^2(\Omega) + b_k^2(\Omega)}. \tag{33–11}$$

Without further refining the distribution law of this variable, we shall show that the mean squared value of its magnitude $|c_k|^2$ is

$$|c_k^2| = \sigma_k^2 = \frac{N-k}{N^2}. \tag{33–12}$$

Using (33–10), we obtain

$$|c_k|^2 = c_k c_k^* = \left(\frac{1}{N} \sum_{i=0}^{N_k - 1} x_i \, e^{j\Omega i \tau_0}\right) \left(\frac{1}{N} \sum_{l=0}^{N-1} x_l \, e^{-j\Omega l \tau_0}\right)$$

$$= \frac{1}{N^2} \sum_i \sum_l x_i x_l \, e^{j(i-l)\Omega\tau_0}.$$

If $l \neq i$ and $k=0$, then x_i and x_l are independent random variables which take on the values ± 1; moreover, the mathematical expectations of x_i and x_l

equal zero. Therefore, in averaging the double sum, there remain only the terms for which $l=i$, and we obtain

$$\overline{|c_k|^2} = \frac{1}{N^2} \sum_{i=0}^{N_k-1} \overline{x_i^2} = \frac{N_k}{N^2} = \frac{N-k}{N^2}.$$

Here $\overline{x_i^2}=1$ is taken into account.

Thus, the RMS value of the ambiguity function side lobes over the entire t, Ω-plane does not exceed $1/\sqrt{N}$; consequently, PM signals with random phase alternations have satisfactory properties from the viewpoint of range and range rate resolution.*

It is also observed from the rough qualitative estimate that the mean level of the side lobes, which is the order of $1/\sqrt{N}$ for the entire t, Ω-plane, is also the minimum possible level. Figure 42 is an "averaged" solid plot of the

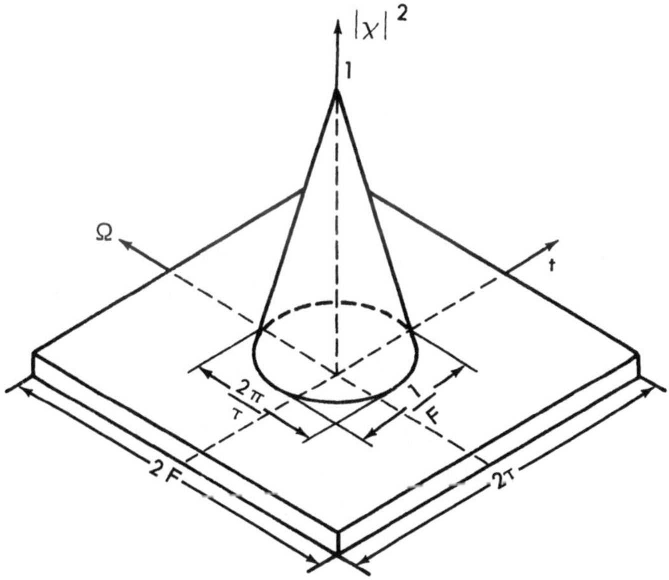

Fig. 42

function $|\chi(t, \Omega)|^2$, which is close to the optimal function. This function has a central peak (cone) whose dimensions at the base are $\Delta t=1/F$ and $\Delta\Omega \approx 2\pi/\tau$ (F is the spectrum width and τ is the signal duration), and a uniform rectangular "side-lobe" region, the level of which is h^2. The dimen-

* The reader is reminded that similar to other signals with a rectangular envelope, the PM pulse has an ambiguity function whose cross section along the frequency axis is

$$\chi(0, \Omega) = \frac{\sin \Omega N \tau_0}{\Omega N \tau_0};$$

thus the highest side lobe for $t=0$ is down by 13 dB.

sions of this region are 2τ and $4\pi F$, respectively. The volume of the cone is given by

$$v_1 = \frac{1}{3}\pi\frac{1}{2F}\frac{2\pi}{2\tau} = \frac{\pi^2}{6F\tau}.$$

The volume of the flat side-lobe region is

$$v_2 = 4\pi F 2\tau h^2 = 8\pi F\tau h^2 .$$

The total volume of the function $|\chi|^2$ equals 2π; therefore,

$$h = \sqrt{\frac{2\pi - v_1}{4\pi F\tau}} \approx \frac{1}{2\sqrt{F\tau}}.$$

Since for PM signals $\tau = N\tau_0$ and $F \approx 1/\tau_0$, we finally have

$$h \approx \frac{1}{2\sqrt{N}}.$$

Thus PM signals having random phase alternations are close to the optimal signal.

[§34] SYNTHESIS OF FM SIGNALS FROM TWO CROSS SECTIONS OF THE AMBIGUITY FUNCTION

In this section we proceed to the synthesis of signals from the ambiguity function over the entire t, Ω-plane. As noted previously, the realizable ambiguity functions belong to a very narrow class of functions. In the general case, it is not possible to find a signal $s(t)$ which realizes any arbitrary function in two variables, $\chi(t, \Omega)$, as an ambiguity function. Such a problem may not have a solution. However, under certain conditions, two cross sections of the ambiguity function are given, as for example, $\chi(t, 0)$ and $\chi(0, \Omega)$, and it is possible to find a signal, whose ambiguity function has such cross sections.

From the Kotel'nikov theorem, a signal of duration T, having the frequency band F, has $N = 2FT$ degrees of freedom. Such a signal is uniquely defined by N arbitrary, independent parameters; in particular, the amplitudes and phases of the spectral components, separated from one another by the interval

$$\Delta\omega = \frac{2\pi}{T},$$

may be chosen as such parameters.

In other words, to determine the signal

$$s(t) = \frac{1}{2\pi} \int_{-\frac{\Omega_m}{2}}^{\frac{\Omega_m}{2}} S(\omega)\, e^{j\omega t}\, d\omega, \tag{34–1}$$

it is sufficient to assign values to $S(\omega_i)$ at the discrete points $\omega_i = i(2\pi/T)$; moreover, two quantities, the magnitude $|S(\omega_i)|$ and the phase $\psi(\omega_i)$, are given at each point, and the number of reference points is $N/2$.

The autocorrelation function

$$R(t) \equiv \chi(t, 0) = \frac{1}{2\pi} \int_{-\frac{\Omega_m}{2}}^{\frac{\Omega_m}{2}} |S(\omega)|^2\, e^{j\omega t}\, d\omega \tag{34–2}$$

cannot have a larger number of degrees of freedom than the signal producing this function. Therefore, $R(t)$ is defined by the values of $S(\omega_i)$ at the above points. However, the autocorrelation function depends only on the magnitude $|S(\omega)|$; the phase structure of the spectrum characterized by $N/2$ parameters has no effect on the form of $R(t)$. Thus, the autocorrelation function of an N-dimensional signal has $N/2$ degrees of freedom. The excess in the signal's degrees of freedom is, in fact, the reason why the signal cannot be fully defined by the autocorrelation function.

From another theorem of Kotel'nikov, the signal

$$s(t) = a(t)\, e^{j\varphi(t)}$$

can be defined by its values $s(t_j)$ at the discrete points

$$t_j = \frac{j}{F} = j\,\frac{2\pi}{\Omega_m}.$$

The number of points is $N/2$; however, two factors, the envelope $a(t_j)$ and the phase $\varphi(t_j)$, are given at each of these points. Again, the total number of parameters equals N, *i.e.*, the number of degrees of freedom of the signal.

Let us now consider the second cross section of the ambiguity function:

$$\chi(0, \Omega) = \int_{-T/2}^{T/2} |s(t)^2|\, e^{j\omega t}\, dt = \int_{-T/2}^{T/2} a^2(t)\, e^{j\omega t}\, dt. \tag{34–3}$$

This cross section depends only on the envelope of the signal and, consequently, is defined by $N/2$ parameters, *i.e.*, by the values of the envelope at the points t_j.

These considerations show that the problem of synthesizing the signal from two cross sections of the ambiguity function, $\chi(t, 0)$ and $\chi(0, \Omega)$, can have only one solution, since these cross sections are given by a total of N parameters. In order for such a solution to exist, the given functions, $\chi(t, 0)$ and $\chi(0, \Omega)$, must satisfy some additional conditions. For the autocorrelation function, such conditions, which were formulated in §24, are as follows: $\chi(t, 0)$ must be an integral function to the power $\Omega_m/2$ and have a positive spectrum; in other words, $\chi(t, 0)$ must belong to the $W^+_{\Omega_m/2}$ class. A comparison between (34–2) and (34–3) makes it possible to formulate the analogous condition for the other cross section: $\chi(0, \Omega)$ must be an integral function to the $T/2$ power belonging* to the $W^+_{T/2}$ class.

Now we can proceed directly to the synthesis of the signal. The method considered below assumes that the FM signal is synthesized with a large compression ratio. This makes it possible to use the results of §31, which are based on the stationary phase method, and this substantially simplifies the problem.

Assume that the basic cross sections of the ambiguity function are given in the form

$$\chi(t, 0) = g(t), \chi(0, \Omega) = h(\Omega),$$

where g and h are integral functions to a finite power. This means that for large complex values of t and Ω, the conditions $|g(t)| \lesssim c_1 e^{p|t|}$ and $|h(\Omega)| \lesssim c_2 e^{q|\Omega|}$ are satisfied. The quantities p and q are designated as the spectrum width and duration of the sought-after signal:

$$\frac{\Omega_m}{2} = p, \quad \frac{T}{2} = q; \tag{34–4}$$

moreover, $T\Omega_m = 4pq \gg 1$. Using the inverse Fourier integrals (34–2) and (34–3) makes it possible to find the power spectrum and the envelope of the signal:

$$\begin{aligned} F(\omega) &= |S(\omega)|^2 = \int_{-\infty}^{\infty} g(t)\, e^{-j\omega t}\, dt, \\ a^2(t) &= \frac{1}{2\pi} \int_{-\infty}^{\infty} h(\Omega)\, e^{-j\omega t}\, d\Omega. \end{aligned} \tag{34–5}$$

Instead of directly evaluating these integrals, one can use the theorem of

* What is meant here is that a signal of finite duration T has a bounded frequency spectrum width F. Such a condition is approximately satisfied for wide-band FM signals (see Appendix 1).

Kotel'nikov. This yields [*cf.* (25–2)]

$$F(\omega)=\frac{1}{2}\sum_{k=-\infty}^{\infty} g\left(\frac{2\pi k}{\Omega_m}\right)e^{-j\frac{2\pi k\omega}{\Omega_m}}, \tag{34–6}$$

$$a^2(t)=\frac{1}{4\pi}\sum_{k=-\infty}^{\infty} h\left(\frac{2\pi k}{T}\right)e^{-j\frac{2\omega kt}{T}}.$$

Since g and h are integral functions to a finite power, the functions $F(\omega)$ and $a(t)$ differ from zero in the intervals $|\omega|\leqslant\Omega_m/2$ and $|t|\leqslant T/2$. If g and h belong to the W^+ class, *i.e.*, have a positive spectrum, then $F(\omega)$ and $a^2(t)$ are positive functions. Otherwise, the magnitudes of the corresponding expressions must be used.* Furthermore, the FM law $\omega_c(t)$ must be determined from the power spectrum $F(\omega)$, and the envelope $a(t)$. As was shown in §31, this law is given by the following integral relationship:

$$\int_0^{\omega_c} F(\omega)\,d\omega = k\int_0^t a^2(t)\,dt, \tag{34–7}$$

where k is a normalization factor.

A signal with an envelope $a(t)$ and instantaneous frequency function $\omega_c(t)$ has an ambiguity function $\chi(t,\Omega)$, whose basic cross sections $\chi(t,0)$ and $\chi(0,\Omega)$ correspond (within the approximation of the stationary phase method) to the given functions $g(t)$ and $h(\Omega)$. It is also possible to use the iteration process given in §31 for refining such a solution.

As an example, let us consider the case in which both cross sections are given by quasi-optimal (Dolph-Chebyshev) functions with the same side-lobe level:

$$g(t)=\frac{\cos\sqrt{(pt)^2-\beta^2}-\cos pt}{M-1},$$

$$h(\Omega)=\frac{\cos\sqrt{(q\Omega)^2-\beta^2}-\cos q\Omega}{M-1},$$

where $\beta=\cosh^{-1}M$.

The factors p and q define the "sharpness" (scale) of the given cross sections, *i.e.*, the resolution of the time and frequency. Then obviously

$$|g(t)|\lesssim\frac{1}{M-1}e^{p|t|},\quad |h(\Omega)|\lesssim\frac{1}{M-1}e^{q|\Omega|};$$

* It follows from §24 that this leads to some improvement in the given functions.

therefore, from (34–4) we obtain

$$\Omega_m = 2p\,, \quad T = 2q\,.$$

The integrals (34–5) were investigated in §27 and in Appendix 4; it was shown that

$$F(\omega) = \frac{I_1\left(\beta\sqrt{1 - \dfrac{4\omega^2}{\Omega_m^2}}\right)}{\sqrt{1 - \dfrac{4\omega^2}{\Omega_m^2}}\, I_1(\beta)}\,, \quad |\omega| < \frac{\Omega_m}{2}\,,$$

$$a^2(t) = \frac{I_1\left(\beta\sqrt{1 - \dfrac{4t^2}{T'^2}}\right)}{\sqrt{1 - \dfrac{4t^2}{T^2}}\, I_1(\beta)}\,, \quad |t| < \frac{T}{2}\,.$$

The graphs of these relationships are given in Figure 32.

Finally, it follows from §31 that (34–7) is satisfied for such $F(\omega)$ and $a(t)$, when the instantaneous frequency changes linearly according to

$$\omega_c(t) = \Omega_m \frac{t}{T}\,, \qquad |t| \leqslant \frac{T}{2}\,.$$

Thus, one of the signals considered in §31 is optimal for the two principal cross sections of the ambiguity function.*

We have considered the synthesis of signals from the principal cross sections of the ambiguity function, $\chi(t, 0)$ and $\chi(0, \Omega)$. Such a problem is analogous, for example, to the synthesis of antennas from the directional pattern in the horizontal and vertical planes. However, these results can be generalized for the cases where the arbitrary ambiguity function cross sections are not along the time and frequency axis. For example, the cross section which is characterized by the angular coefficient γ_1 that corresponds to the equation

$$\Omega = \gamma_1 t\,,$$

has the following ambiguity function:

$$\chi(t, \gamma_1 t) = \chi_1(t) = \int_{-\infty}^{\infty} s\left(t' + \frac{t}{2}\right) s^*\left(t' - \frac{t}{2}\right) e^{j\gamma_1 t t'}\, dt'\,. \qquad (34\text{–}8)$$

* An equivalent synthesis method is proposed in [32]. Although the problem has a different approach in this work from that given here, it can be shown that both methods are only suitable for wide-band FM signals whose frequency changes monotonically (this fact is not considered in [32]). Also, the order of accuracy of both the methods is the same.

Let us consider a new signal $s_1(t)$ which is related to $s(t)$ by

$$s(t) = s_1(t)\, \mathrm{e}^{-j\frac{\gamma_1}{2}t^2}. \tag{34–9}$$

Substituting this value into (34–8), we find

$$\begin{aligned}\chi_1(t) &= \int_{-\infty}^{\infty} s_1\left(t' + \frac{t}{2}\right) s_1^*\left(t' - \frac{t}{2}\right) \mathrm{e}^{j\gamma_1\left[tt' - \frac{1}{2}\left(t' - \frac{t}{2}\right)^2 + \frac{1}{2}\left(t' - \frac{t}{2}\right)^2\right]}\, dt' \\ &= \int_{-\infty}^{\infty} s_1\left(t' + \frac{t}{2}\right) s_1^*\left(t' - \frac{t}{2}\right) dt'.\end{aligned}$$

Hence, it is clear that the autocorrelation function of the signal $s_1(t)$ coincides with the cross section of the autocorrelation function of the signal $s(t)$ under consideration. Noting that both signals have the same envelope and, therefore, the same principal cross section $\chi(0, \Omega)$, we arrive at the following result.

In order to synthesize the signal $s(t)$ from the principal cross section $\chi(0, \Omega)$, and an arbitrary cross section with angular coefficient γ_1, it is necessary to determine beforehand the signal $s_1(t)$ for which both given functions are the principal cross sections and then to find the sought-after signal using (34–9).

Similarly, it is not difficult to show that the signal $s_2(t)$ whose spectrum is defined by

$$S_2(\omega) = S(\omega)\, \mathrm{e}^{-j\frac{\omega^2}{2\gamma_2}}, \tag{34–10}$$

has an ambiguity function whose principal cross section $\chi_2(0, \Omega)$ coincides with the cross section of the initial signal's ambiguity function; the angular coefficient γ_2 characterizes this cross section, *i.e.*,

$$\Omega = \gamma_2 t.$$

Since the power spectra of the signals $s(t)$ and $s_2(t)$ are the same, it is possible to synthesize the signals by the autocorrelation function $\chi(t, 0)$ and by any other cross section of the ambiguity function.

Combining both results, we arrive at the conclusion that it is possible to synthesize a signal from two cross sections of the ambiguity function, one of which must be a principal cross section. In the next section we shall prove the rotation theorem which allows one (principal) cross section of the ambiguity function to be placed at any arbitrary angle.

[§35] ROTATION OF THE AMBIGUITY FUNCTION

The method of the previous section makes it possible to synthesize a signal from two cross sections of the ambiguity function. Also the synthesis potentialities were largely exhausted since, as was shown, the number of degrees of freedom of the signal exactly corresponds to the number of independent parameters which define the cross sections.

If the ambiguity function is given with more completeness, as for example, by three cross sections, or even over the entire t, Ω-plane, the synthesis problem, as a rule, does not have an exact solution. For such a situation, we can speak either of an approximate synthesis (where the degree of approximation depends mostly on the given function), or of special functions possessing certain characteristic properties. This is, in fact, the topic of this section; the method of approximate synthesis is discussed in §37 and §38.

Let the signal $s(t)$, which realizes the ambiguity function

$$\chi_1(t, \Omega) = \frac{1}{E_1} \int_{-\infty}^{\infty} s_1\left(t' + \frac{t}{2}\right) s_1^*\left(t' - \frac{t}{2}\right) e^{j\Omega t'} \, dt,$$

be subjected to the transformation $L(s)$ that results in the formation of another signal

$$s_2(t) = L[s_1(t)].$$

This new signal has the ambiguity function

$$\chi_2(t, \Omega) = \frac{1}{E_2} \int_{-\infty}^{\infty} s_2\left(t' + \frac{t}{2}\right) s_2^*\left(t' - \frac{t}{2}\right) e^{j\Omega t'} \, dt',$$

which can be called the *realizable deformation* of $\chi_1(t, \Omega)$ if the transformation $L(s_1)$ can be performed. (Of course, the discussion here is theoretical and does not consider the engineering problem.)

Let us consider a transformation of the type

$$s_2(t) = \sqrt{\frac{b}{2\pi}} \int_{-\infty}^{\infty} s_1(x) \, e^{j\left[\frac{a}{2}(x^2+t^2) - bxt\right]} \, dx. \tag{35–1}$$

Accordingly, for the ambiguity function we have

$$\chi_2(t, \Omega) = \frac{b}{2\pi E_2^2} \iiint_{-\infty}^{\infty} s_1(x) \, s_1^*(y) \, e^{j(x+y)\left[\frac{a}{2}(x-y) - b\frac{t}{2}\right]}$$
$$\times \, e^{-jb\left[(x-y) - \frac{b}{a}t - \frac{\Omega}{b}\right]t'} \, dx \, dy \, dt'.$$

Changing the order of integration and using expression

$$\int_{-\infty}^{\infty} e^{-jb\left(x-y-\frac{a}{b}t-\frac{\Omega}{b}\right)t'} dt' = \frac{2\pi}{b}\,\delta\left(x-y-\frac{a}{b}t-\frac{\Omega}{b}\right),$$

we obtain

$$\chi_2(t,\Omega) = \frac{1}{E_2}\int_{-\infty}^{\infty} s_1\left(y+\frac{a}{b}t+\frac{\Omega}{b}\right) s_1^*(y)\, e^{j\left(2y+\frac{a}{b}t+\frac{\Omega}{b}\right)\left[\frac{a}{2}\left(\frac{a}{b}t+\frac{\Omega}{b}\right)-b\frac{t}{2}\right]}\, dy.$$

Let us make a substitution of variables, *i.e.*, let

$$y_1 = y + \frac{1}{2}\left(\frac{a}{b}t+\frac{\Omega}{b}\right),$$

and introduce the notation

$$t_1 = \frac{a}{b}t + \frac{1}{b}\Omega, \quad \Omega_1 = \frac{a^2-b^2}{b}t + \frac{a}{b}\Omega. \tag{35–2}$$

In addition, let us determine the energy E_2 with $\chi_2(0,0)=1$. Finally, we obtain

$$\chi_2(t,\Omega) = \frac{1}{E_1}\int_{-\infty}^{\infty} s_1\left(y_1+\frac{t_1}{2}\right) \times s_1^*\left(y_1-\frac{t_1}{2}\right) e^{jy_1\Omega}\, dy_1 = \chi_1(t_1,\Omega_1). \tag{35–3}$$

This equation expresses a remarkable result. It shows that the transformation (35–1) converts the signal $s_1(t)$ into the signal $s_2(t)$ such that the ambiguity function $\chi_2(t,\Omega)$ is exactly the same as the initial function $\chi_1(t,\Omega)$, if the latter is subjected to the linear transformation (35–2). In order to clarify the meaning of this transformation, let us introduce the notation

$$\frac{a}{b} = \cos\theta, \quad b = \frac{1}{\tau^2\sin\theta},$$

where τ is an arbitrary scale unit of time.

Then, (35–2) takes the form

$$\begin{aligned} \frac{t_1}{\tau} &= \frac{t}{\tau}\cos\theta + (\Omega\tau)\sin\theta, \\ \frac{\Omega_1}{\tau} &= -\frac{t}{\tau}\sin\theta + (\Omega\tau)\cos\theta. \end{aligned} \tag{35–4}$$

It is easy to see that in the dimensionless coordinates, t/τ and $\Omega\tau$, the transformation reduces to a rotation of the coordinate axes by the angle θ. Consequently, by changing θ and τ, it is possible to rotate arbitrarily the ambiguity function and to change simultaneously the time and frequency scales so that they remain matched.

It is pointed out that by using the new notation, the transformation (rotation) (35–1) has the form

$$s_2(t) = \frac{1}{\tau}\sqrt{\frac{\csc\theta}{2\pi}} \int_{-\infty}^{\infty} s_1(x)\, e^{j\left(\frac{x^2+t^2}{2\tau^2}\cot\theta - \frac{xt}{\tau^2}\csc\theta\right)}\, dx. \qquad (35\text{–}5)$$

The transformation is given in this form by Klauder [33]. From the viewpoint of synthesis, this transformation is of interest because it makes possible the design of a signal whose ambiguity function $\chi_1(t, \Omega)$ differs from the known realizable function by a rotation in the t, Ω-plane.

The transition from a simple AM signal to an FM pulse may serve as an interesting application of the above transformation. Let $s_1(t)$ be an arbitrary amplitude envelope of a signal having a sufficiently long duration T. This only means that $s_1(t)$ changes noticeably over a time interval the order of T. Let us choose the scale unit τ of (35–5) in such a manner that $\tau \ll T$. In addition, we shall also assume that the angle θ is restricted by $0<\theta<\pi/2$, and that θ is not too close to the boundaries of this interval. Then, rewriting (35–5) in the form

$$s_2(t) = \frac{1}{\tau}\sqrt{\frac{\csc\theta}{2\pi}}\, e^{j\frac{t^2}{2\tau^2}\cot\theta} \int_{-\infty}^{\infty} s_1(x)\, e^{j\frac{\csc\theta}{\tau^2}\left(\frac{x^2}{2}\cos\theta - xt\right)}\, dx,$$

we can use the stationary phase method to evaluate the integral approximately. In fact, for the assumptions made, $s_1(t)$ is a slowly varying function as compared to the exponential multiplier. The point of stationary phase, x_0, is determined from condition

$$\frac{d}{dx}\left(\frac{x^2}{2}\cos\theta - xt\right)\bigg|_{x=x_0} = x_0\cos\theta - t = 0.$$

Therefore, $x_0 = t\sec\theta$. Furthermore, by using the expressions from the stationary phase method ([9], p. 19; also, see Appendix 1), we obtain, after some simple transformations,

$$s_2(t) = \frac{s_1(t\sec\theta)}{\sqrt{\cos\theta}}\, e^{-j\left(\frac{t^2}{2\tau^2}\cot\theta - \frac{\pi}{4}\right)}. \qquad (35\text{–}6)$$

It is clear that $s_2(t)$ is a signal whose frequency changes linearly. The envelope of this signal is the same as the envelope of the initial signal

$s_1(t)$, but with a different time scale. Thus, introducing sufficiently steep linear FM into a signal with any arbitrary, slowly changing envelope, leads (in the first approximation) to a rotation of the ambiguity function in the t, Ω-plane. This result is a generalization of a special case of the "Gaussian" signal considered in §9.

Another important result based on this transformation (rotation), consists of the determination of signals whose ambiguity functions have circular symmetry, *i.e.*, they are figures of revolution. For example, in the cases where the measurement of range and range rate are equally important and it is also not possible to specify any directional advantage in the t, Ω-plane, the ambiguity function with circular symmetry must be considered as preferable. The above transformation (rotation) makes it possible to determine signals with such properties.

First of all, let us note that a "Gaussian" signal with the envelope

$$s(t) = \mathrm{e}^{-t^2/T^2}$$

has such properties. As was shown in §9, its ambiguity function

$$\chi(t, \Omega) = \mathrm{e}^{-\frac{1}{2}\left(\frac{t^2}{T^2} + \frac{\Omega^2}{(2/T)^2}\right)}$$

has an elliptical form in the horizontal direction. If, however, the dimensionless coordinates

$$\eta = \frac{t}{T}, \quad \xi = \frac{\Omega T}{2} \tag{35–7}$$

are introduced, then this function becomes a figure of revolution,

$$\chi(\eta, \xi) = \mathrm{e}^{-\frac{1}{2}(\eta^2 + \xi^2)},$$

and the height profiles are circles. In order to eliminate the effect of the time and frequency scales, we will use the dimensionless variables of (35–7) where T is an arbitrary time unit (instead of t and Ω). The circular symmetry also holds for these coordinates. It is clear that by going back to the variables t and Ω, it is possible to deform the circles into ellipses with an arbitrary ratio of the axes.

In the new coordinates (35–7), Eqs. (35–4) and (35–5) take the form

$$\left.\begin{aligned} k\eta_1 &= k\eta \cos\theta + \frac{\xi}{k} \sin\theta, \\ \frac{\xi_1}{k} &= -\,k\eta \sin\theta + \frac{\xi}{k} \cos\theta, \end{aligned}\right\} \tag{35–8}$$

$$s_2(\eta) = k\sqrt{\frac{\csc\theta}{2\pi}} \int_{-\infty}^{\infty} s_1(\eta')\, \mathrm{e}^{jk^2\left(\frac{\eta_1{}^2 + \eta^2}{2}\cot\theta - \eta\eta' \csc\theta\right)}\, d\eta', \tag{35–9}$$

where $k = T/\tau$.

The ambiguity function $\chi(\eta, \xi)$ has circular symmetry if in substituting η_1 and ξ_1 for η and ξ with $k=1$, equation

$$\chi(\eta, \xi) = \chi(\eta_1, \xi_1)$$

is valid for any arbitrary θ.

It follows from §23 that if the arbitrary initial phase is not taken into account, the realizable ambiguity function uniquely defines the signal form. Therefore, signals which realize a function with circular symmetry must satisfy

$$\sqrt{\frac{\csc\theta}{2\pi}} \int_{-\infty}^{\infty} s(\eta')\, e^{j\left(\frac{\eta'^2+\eta^2}{2}\cot\theta - \eta\eta'\csc\theta\right)} d\eta' = s(\eta)\, e^{-j\varphi}, \tag{35–10}$$

which is obtained from (35–9) by requiring that condition $S_2(\eta)=S_1(\eta)\,e^{-j\varphi}$ be satisfied for any arbitrary θ. Signals which realize ambiguity functions with circular symmetry must be invariant (not considering the initial phase) with respect to the transformation (rotation). It is noted that the phase φ depends on θ, since complete invariance is only possible in the trivial case of $S(\eta)\equiv 0$.

The set of solutions (eigenfunctions) of the integral equation (35–10) can be found in the following manner. Introducing the notation

$$z = e^{-j\theta},$$

and making use of

$$\cot\theta = j\,\frac{1+z^2}{1-z^2}, \quad \csc\theta = \frac{2jz}{1-z^2},$$

we can rewrite (35–10) in the form

$$s(\eta) = \frac{\lambda}{\sqrt{2\pi(1-z^2)}} \int_{-\infty}^{\infty} s(\eta)\, e^{-\left(\frac{1+z^2}{1-z^2}\frac{\eta'^2+\eta^2}{2} - \frac{2z}{1-z^2}\eta\eta'\right)} d\eta', \tag{35–11}$$

where

$$\lambda = \sqrt{jz}\, e^{j\varphi} = e^{j\left(\varphi - \frac{\theta}{2} + \frac{\pi}{4}\right)}. \tag{35–12}$$

It is known [65] that the eigenfunctions of (35–11) are Hermitian functions:

$$s_n(\eta) = D_n(\sqrt{2}\eta) = \frac{1}{\sqrt{2^n \pi^{1/2} n!}}\, e^{-\frac{\eta^2}{2}} H_n(\eta). \tag{35–13}$$

For every eigenfunction $s_n(\eta)$, there is a corresponding eigenvalue λ_n, defined by

$$\lambda_n = z^{-n} = e^{jn\theta}.$$

Comparing this value with (35–12), we obtain

$$\varphi_n = (n + \tfrac{1}{2})\,\theta - \frac{\pi}{4}. \tag{35–14}$$

Thus, the Hermitian functions represent signals (more accurately, the signal envelopes) which result in ambiguity functions with circular symmetry. The case of $n=0$ yields a Gaussian pulse.

Proceeding to the determination of the corresponding ambiguity functions, we find, considering (35–13),

$$\begin{aligned}\chi(\eta, \xi) &= \frac{1}{2^n \sqrt{\pi n!}} \int_{-\infty}^{\infty} e^{-\frac{(\eta' + \eta/2)^2 + (\eta' - \eta/2)^2}{2}} H_n(\eta' + \eta/2)\, H_n(\eta' - \eta/2)\, e^{j\xi\eta'}\, d\eta' \\ &= \frac{e^{-\eta^2/4}}{2^n \sqrt{\pi n!}} \int_{-\infty}^{\infty} e^{-\eta'^2 + j\xi\eta'} H_n(\eta' + \eta/2)\, H_n(\eta' - \eta/2)\, d\eta \\ &= \frac{e^{-\frac{\eta^2+\xi^2}{4}}}{2^n \sqrt{\pi n!}} \int_{-\infty}^{\infty} e^{-x^2} H_n\!\left(x + \frac{\eta + j\xi}{2}\right) H_n\!\left(x - \frac{\eta - j\xi}{2}\right) dx \\ &= e^{-\frac{\eta^2+\xi^2}{4}} L_n\!\left(\frac{\eta^2 + \xi^2}{2}\right) = e^{-r^2/4} L_n(r^2/2),\end{aligned} \tag{35–15}$$

where $r^2 = \eta^2 + \xi^2$ and L_n is the Laguerre polynomial. In the above calculations, the value of the known integral with Hermitian polynomials is considered (Gradshtein & Ryzhik [17], p. 836).

Figure 43 shows the graphs (from [33]) of the signal envelopes $s_n(\eta)$ and their ambiguity functions for n = 0, 4, 10. All the curves are symmetrical and thus are only plotted for the positive arguments. It is clear from these graphs that ambiguity functions with circular symmetry have comparatively large and slowly decreasing side lobes. The maximum side-lobe level is relatively independent of n and is approximately 30% of the maximum. Consequently, in order to obtain ambiguity functions with circular symmetry, one must sacrifice both side-lobe level and the rate of their reduction.

The envelopes of the $s_n(\eta)$ signals are alternating. This means that the phase of the carrier changes by π. Consequently, the signals which realize symmetrical ambiguity functions are close to PM pulses. The main differences lie in the fact that in addition to changes in the phase of the carrier there are also amplitude changes.

It is also interesting to note that at least for large n, the signal amplitude increases towards the edges of the pulse. As is known, the Hermitian func-

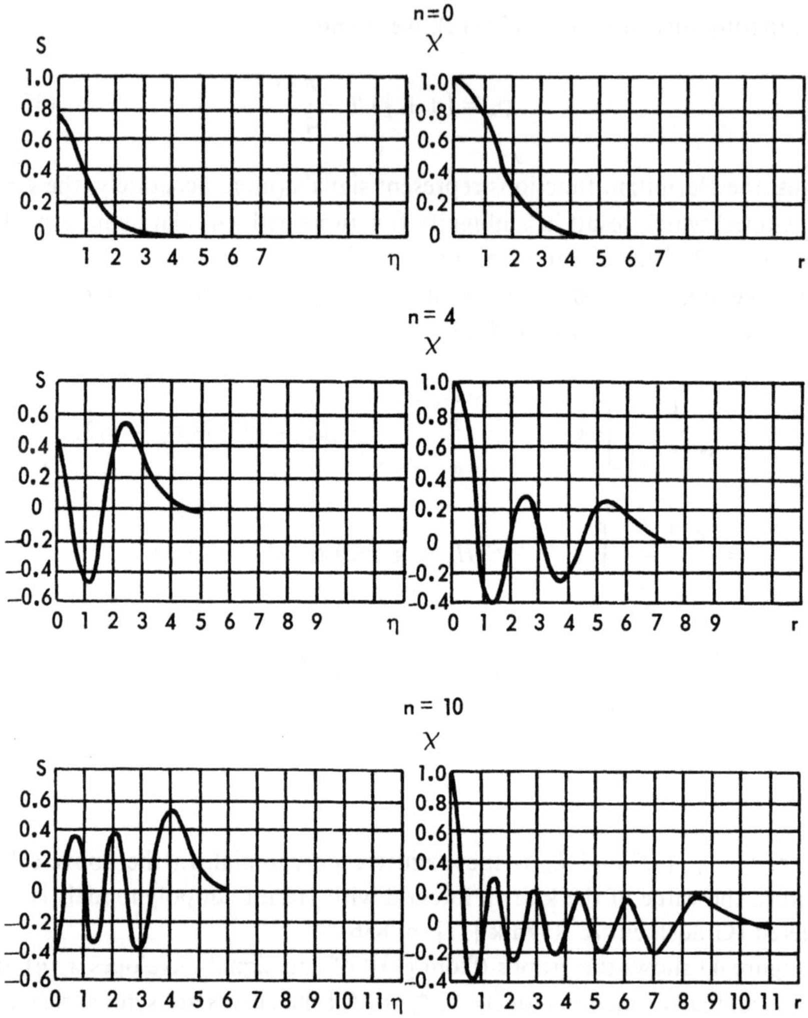

Fig. 43

tions are self-reciprocal for the Fourier transform. This means that the spectra of the signals $s_n(\eta)$ are represented by the same function as the signal itself. [This explains why the ambiguity function cross sections $\chi(\eta, 0)$ and $\chi(0, \xi)$ are identical.] Consequently, the spectral functions of the signals also increase towards the edges of the frequency band. Here we have the same analogy with the signals which are optimal in the Dolph-Chebyshev sense. As noted in §27, the increase of spectral density at the edges of the frequency band is characteristic of signals which are optimal in the above sense. For the case being considered, this is not only true of the spectrum, but also of the signal which is a function of time.

In addition, it follows from Appendix 5 that the Hermitian functions are close to the spheroidal functions. The corresponding spectra have the same properties, since the spheroidal functions are also approximately self-reciprocal for the Fourier transform. Therefore, the signals $s_n(\eta)$, which form the circularly symmetrical ambiguity functions, are not only related to the Dolph-Chebyshev signals, but also to those which are optimum in the sense of the quadratic approximation (*cf.* §28).

[§36] FM SIGNALS WITH UNIFORM SIDE-LOBES IN THE t, Ω-PLANE

It has been seen from our analysis that the synthesis of FM signals from autocorrelation functions can be substantially simplified by using the stationary phase method. This method readily incorporates the basic requirements used in choosing the signals. Such a solution has acceptable accuracy from the engineering viewpoint and provides for the clarity of calculation. The application of this method is associated with the characteristic FM signals used in radar, *i.e.*, wide-band signals which are obtained at the expense of large frequency modulation. It is natural to use this property for a more general synthesis problem and to apply the stationary phase method for constructing FM signals when considering the side lobes in the entire t, Ω-plane.

The ideal ambiguity function has one central peak and has its side lobes distributed uniformly in a rectangular region having the dimensions $2T$, $2F$ (see Fig. 42). It was noted previously that random PM pulses have approximately such properties. FM signals with a monotonic (in particular linear) frequency characteristic are comparatively far from being ideal. The center "peak" of any ambiguity function from these signals is elongated in the t, Ω-plane. This leads to poor target discrimination when the range and range rate are linearly related. By observing the changes of $|\chi(t, \Omega)|$ as a function of Doppler frequency, Ω, for fixed values of t, it can be seen (as shown in Fig. 20) that the "side lobes" change noticeably with Ω. One approximation to the optimal ambiguity function is the case in which the side-lobe level is independent of Ω, *i.e.*, in which the condition

$$|\chi(t, \Omega)| = c(t) \tag{36–1}$$

is satisfied outside a specified center region or, stated in another way, this relationship is

$$\frac{\partial |\chi(t, \Omega)|}{\partial \Omega} = 0 . \tag{36–2}$$

It will be shown in the following that FM signals have approximately such properties, but in this case the frequency law is not monotonic.

First, let us consider the case of an FM signal with a rectangular envelope:

$$s(t) = e^{j\varphi(t)} \quad \text{for} \quad -\frac{T}{2} \leqslant t \leqslant \frac{T}{2}.$$

The ambiguity function of this signal is

$$\chi(t, \Omega) = \frac{1}{T} \int_{-\frac{T-|t|}{2}}^{\frac{T-|t|}{2}} e^{j\left[\varphi\left(t'+\frac{t}{2}\right)-\varphi\left(t'-\frac{t}{2}\right)+\Omega t'\right]} dt'. \tag{36–3}$$

From the stationary phase method, the value of this integral is defined mainly in the region of the stationary point t_0, which satisfies the condition

$$\varphi'\left(t_0 + \frac{t}{2}\right) - \varphi'\left(t_0 - \frac{t}{2}\right) + \Omega = 0.$$

Introducing the notation

$$\theta(t', t) = \varphi\left(t' + \frac{t}{2}\right) - \varphi\left(t' - \frac{t}{2}\right), \tag{36–4}$$

the last equation takes the form

$$\theta'(t_0, t) + \Omega = 0, \tag{36–5}$$

where the derivative is taken with respect to t_0.

Assuming that there is only one stationary point, t_0 (in the following, the validity of this assumption will be discussed, as well as some of the other assumptions associated with the stationary phase method), we obtain for the ambiguity function

$$\chi(t, \Omega) \approx \frac{1}{T}\sqrt{\frac{2\pi}{|\theta''(t_0, t)|}}\, e^{j\left[\theta(t_0, t) + \Omega t_0 \pm \frac{\pi}{4}\right]},$$

and, consequently,

$$|\chi(t, \Omega)| = \frac{1}{T}\sqrt{\frac{2\pi}{|\theta''(t_0, t)|}}. \tag{36–6}$$

It can be seen from (36–5) that the stationary point t_0 is a function of the Doppler frequency Ω. Therefore, (36–1) which expresses the independence of the magnitude $|\chi(t, \Omega)|$ from the Doppler frequency reduces to the requirement of $\theta''(t_0, t)$ being independent of t_0, *i.e.*,

$$\theta''(t_0, t) = c(t).$$

Integrating with respect to t_0 leads to relationship

$$\theta(t, t_0) = c_0 + c_1 t_0 + c_2 t_0^2, \tag{36–7}$$

where c_0, c_1, and c_2 can be arbitrary functions of t.

Let the signal's phase modulation function $\varphi(t)$ be an n-th degree polynomial with constant coefficients:

$$\varphi(t) = a_0 + a_1 t + \cdots + a_n t^n.$$

It follows from (36–4) that in this case $\theta(t_0, t)$ is an $(n-1)$-degree polynomial with respect to the variable t_0, *i.e.*,

$$\theta(t_0, t) = b_0 + b_1 t_0 + \cdots + b_{n-1} t_0^{n-1};$$

moreover, the coefficients b_0, b_1, etc., are polynomials of the variable t. By comparing the last relationship with (36–7), it is seen that the power n must not be higher than three. Consequently, in the approximation using the stationary phase method, condition (36–1) is satisfied if the signal's phase modulation law is an arbitrary third-degree polynomial, *i.e.*,

$$\varphi(t) = a_0 + a_1 t + a_2 t^2 + a_3 t^3. \tag{36–8}$$

In this case, θ is a quadratic function and, as can be readily verified, the integral (36–3) cannot contain more than one stationary point. Consequently, the assumption on the uniqueness of the stationary point is justified.

The initial phase a_0 and the linear term $a_1 t$ do not affect the ambiguity function. The quadratic term $a_2 t^2$ corresponds to the linear change of the instantaneous frequency and, as pointed out previously, leads to a "stretching" of the ambiguity function's center peak. Therefore, the quadratic term exerts a harmful effect, and it is expedient to retain only the cubic term of (36–8), *i.e.*, let

$$\varphi(t) = a t^3. \tag{36–9}$$

The coefficient a can easily be expressed in terms of the deviation of the frequency and the pulse duration. The cubic phase modulation law obviously corresponds to the quadratic instantaneous frequency law

$$\omega_0(t) = 3at^2,$$

which is symmetrical about $t=0$. The maximum frequency deviation occurs at the pulse cutoff where $t = \pm T/2$, *i.e.*,

$$\omega_d = 3a\,\frac{T^2}{4}.$$

Let us also introduce the expression for the compression ratio:

$$m = \frac{\omega_d T}{2\pi}.$$

Then (36–9) can be written in the form

$$\varphi(t) = \frac{8\pi m}{3}\frac{t^3}{T^3}. \tag{36–10}$$

From (36–3), the ambiguity function of a signal with cubic phase modulation has the form

$$\chi(t, \Omega) = \frac{e^{j\frac{2}{3}\pi m \frac{t^3}{T^3}}}{T} \int\limits_{-\frac{T-|t|}{2}}^{\frac{T-|t|}{2}} e^{j\left(2\pi m \frac{tt'^2}{T^3} + \Omega t'\right)} dt',$$

or using the dimensionless coordinates

$$\eta = \frac{t}{T}, \quad \xi = \frac{\Omega}{\omega_d} = \frac{\Omega T}{2\pi m},$$

we have

$$\chi(\eta, \xi) = e^{j\frac{2}{3}\pi m \eta^3} \int\limits_{-\frac{1-|\eta|}{2}}^{\frac{1-|\eta|}{2}} e^{j2\pi m(4\eta\eta'^2 + \xi\eta')} d\eta'. \tag{36–11}$$

It is not difficult to reduce this expression to Fresnel integrals which are tabulated, but the result is cumbersome for making computations. It is considerably simpler to calculate the approximate ambiguity function using, once again, the stationary phase method. The stationary point η_0 is determined from condition

$$8\eta\eta_0 + \xi = 0,$$

i.e., $\eta_0 = -\xi/8\eta$. In addition, $\theta''(\eta_0) = 8\eta$ and $\theta(\eta_0) = -\xi^2/8\eta$.

Therefore, in the approximation using the stationary phase method, we obtain

$$\chi(\eta, \xi) \approx \frac{1}{\sqrt{8m|\eta|}} e^{j\pi m\left(\frac{2}{3}\eta^3 - \frac{\xi^2}{4\eta}\right)}. \tag{36–12}$$

Thus, in the side-lobe region, the magnitude of the ambiguity function does not depend on the Doppler frequency Ω, and decreases with time as $1/\sqrt{t}$. The approximate expression (36–12) is only applicable to a certain portion of the t, Ω-plane; in particular, it is suitable in the neighborhood of the principal maximum as $\eta \to 0$. In order to define more clearly the region of applicability, let us consider the conditions under which the stationary phase method can be applied to integral (36–11).

The stationary phase point η_0 must, obviously, lie inside the integration interval, *i.e.*, the condition

$$-\frac{1-|\eta|}{2} \leqslant \eta_0 \leqslant \frac{1-|\eta|}{2}$$

must be satisfied. Considering that $\eta_0 = -\xi/8\eta$, we obtain

$$-4|\eta|(1-|\eta|) \leqslant \xi \leqslant 4|\eta|(1-|\eta|). \qquad (36\text{–}13)$$

It can be readily shown that this condition is satisfied inside the regions bounded by the segments of a parabola (see Fig. 44) which enclose a substantial portion of the side lobes. However, these regions do not contain the

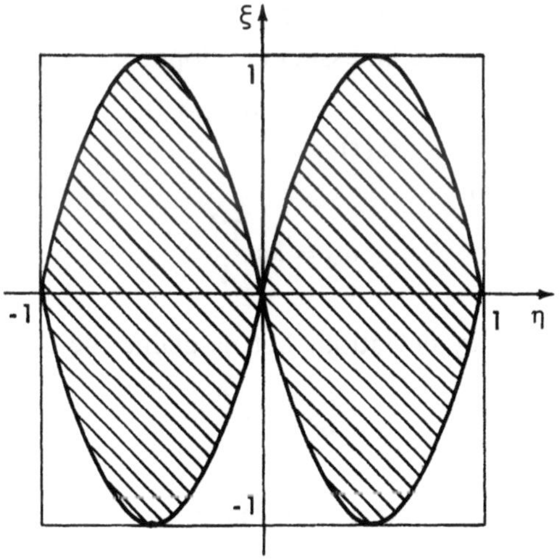

Fig. 44

central peak nor the cross section of $\eta = 0$. In fact, for a signal with a rectangular envelope, the cross section of the ambiguity function along the frequency axis has the form $(\sin \xi)/\xi$.

In order to be able to use the stationary phase method, it is also necessary that the stationary point be not too close to the boundaries of the integration interval ([9], p. 69). As applied to the integral of (36–11), this requirement has the form

$$|4(1-|\eta|)\,\eta^2 - \xi|\eta|| \geqslant \frac{1}{\pi m}.$$

This condition somewhat strengthens requirement (36–13): the approximate expression (36–12) can be used inside the regions shown in Figure 44, but not too close to its boundaries. The minimum distance from the boundary depends on the compression ratio m. In particular, for the autocorrelation function (with $\xi=0$) we have the estimate

$$\frac{1}{\sqrt{2\pi m}} \ll |\eta| < 1 .$$

This inequality shows that by using the stationary phase method, it is only possible to evaluate side lobes which are far from the principal maximum; for the close-in side lobes (as well as for the center region), this method is not suitable.

The above difficulties for estimating the ambiguity functions of signals having a quadratic frequency law are related to the fact that we have considered pulses of finite duration, and the integral of (36–11) has finite limits. In this connection, let us next consider a signal having the same frequency modulation law, but with a Gaussian envelope:

$$s(t) = e^{-\frac{t^2}{T^2}} e^{jat^3} . \tag{36–14}$$

As before, the function

$$\omega_d = 3a \frac{T^2}{4}$$

is used as the frequency deviation. Let us define the compression ratio as

$$m = \frac{\omega_d T}{2\pi},$$

and introduce the dimensionless coordinates

$$\eta = \frac{t}{T}, \quad \xi = \frac{\Omega}{\omega_d} = \frac{\Omega T}{2\pi m} .$$

Then the following expression is obtained for the ambiguity function:

$$\chi(\eta, \xi) = \sqrt{\frac{2}{\pi}}\, e^{-\frac{\eta^2}{2} + j\frac{2}{3}\pi m\eta^3} \int_{-\infty}^{\infty} e^{-2\eta'^2 + j2\pi m(4\eta\eta'^2 + \xi\eta')}\, d\eta' .$$

Completing the square in the exponent, we find

$$|\chi(\eta, \xi)| = \frac{1}{\sqrt[4]{1 + 16\pi^2 m^2 \eta^2}}\, e^{-\frac{1}{2}\left(\eta^2 + \pi^2 m^2 \frac{\xi^2}{1 + 16\pi^2 m^2 \eta^2}\right)} . \tag{36–15}$$

The properties of such an ambiguity function are quite satisfactory. For small η, the magnitude $|\chi(\eta, \Omega)|$ varies as $e^{-\frac{\pi^2 m^2}{2}\xi^2}$. Consequently, the peak along the frequency axis is sufficiently sharp. The width of this peak decreases substantially with an increase of the compression ratio, m. For large η, the ambiguity function level decreases as $1/\sqrt{4\pi m\eta}$, and the side lobes become smaller as the compression ratio m increases. In this region, the side-lobe levels are weakly dependent on frequency, since the dependence on m vanishes in the exponential factor. The characteristic dimensions of the center peak are given by

$$\Delta\eta \sim \frac{1}{4\pi m}, \quad \Delta\xi \sim \frac{1}{\pi m}$$

or

$$\Delta t \sim \frac{T}{4\pi m} = \frac{1}{2\omega_d}, \quad \Delta\Omega \sim \frac{\omega_d}{\pi m} = \frac{2}{T}.$$

The mean level of the side lobes is estimated by

$$|\Delta\chi| \sim \frac{1}{\sqrt{2\pi m}}.$$

Thus, signals having a quadratic frequency law are close, with respect to the ambiguity function parameters, to the random PM signals and to the "ideal" signals with uniform side lobes.

Signals with non-monotonic (quadratic, Gaussian) frequency laws are not only of theoretical but also of practical interest. A radar system having good range and range rate resolutions, in which signals similar to those considered in this section are used, is described in the literature [51]. Two pulses with linear frequency laws are emitted by this system; however, for one of the pulses, the frequency increases with time, and in the other pulse, the frequency decreases with time (see Fig. 45a). Such a piecewise linear, frequency modulation law is close to the quadratic law. Each pulse with linear FM is received with the corresponding matched filter at the input. As a result, two compressed pulses are reflected from a stationary target. For pulses with linear FM, the changes in the Doppler frequency, as well as in the delay time, lead to the compressed signals being shifted in time. In the case of a moving target, both compressed signals are shifted on the swept indication, but due to the different signs of the FM laws, these shifts occur in different directions. Therefore, the value of the target's range rate can be established from the interval between the compressed pulses, and the target range from a pair of pulses on the indicator (see Fig. 45b). Although such a processing system differs from optimal, the potentiality of using such signals for the simultaneous measurement of range and range rate qualitatively confirms the properties of the ambiguity function considered in this section.

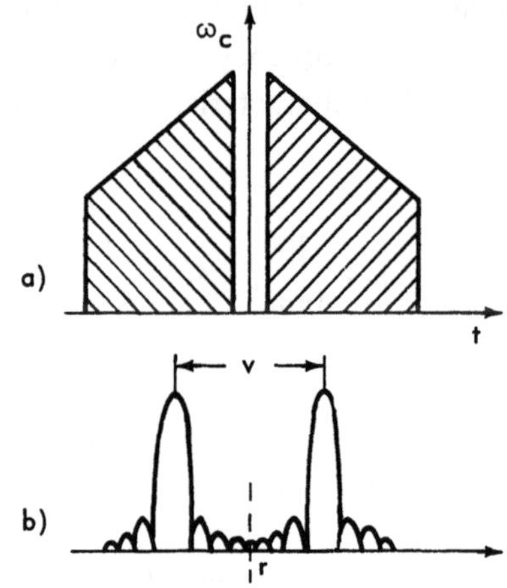

Fig. 45

It is also noted that by using the "key-lock" method it is possible to construct such a system very economically. It was demonstrated in §7 that the optimal filter shown in Figure 13 can be used for both the formation and reception processing of pulses having linear FM. If a short pulse (δ-function) is supplied as an input to this filter, then an FM signal will be formed at the output. The received reflected signal must be supplied to another filter input to form a compressed pulse at the output. It is important that the system under consideration be capable of supplying FM signals with both increasing and decreasing frequency laws from δ-functions supplied to various filter inputs. Therefore, it is possible to construct a system for the formation and processing of signals in the following manner.

For the first transmission, the exciting δ-function is supplied to Input I in Figure 13, and the reception of this signal is accomplished at Input II. Conversely, for the second transmission, Input II is used for the formation, while Input I is used for its reception. Furthermore, the process is repetitive; as a result, only one pulse with linear FM is generated during each transmission, but the direction of the FM is changed from period to period. When the received signals coincide on the indicator, the picture shown in Figure 45b is obtained. Thus, only one matched filter is used for the formation and processing system.

[§37] APPROXIMATE SYNTHESIS METHOD FROM THE MAGNITUDE AND PHASE OF AN ARBITRARY AMBIGUITY FUNCTION

It could be assumed that the synthesis methods for FM, PM, and those signals considered here for which $\chi(t, \Omega)$ has circular symmetry, have practically exhausted the need in the radar signal synthesis field. In fact, the first two groups of methods include the class of signals which can be realized simply by appropriate devices; the last method yields optimal solutions which makes it possible to design systems having the best properties with respect to the simultaneous resolution of range and range rate. In spite of this, we shall consider a general synthesis method from the ambiguity function given over the entire t, Ω-plane. This method, proposed by Sussman [56], is undoubtedly of importance from the theoretical viewpoint, since it makes possible the rigorous restriction of the class of functions which can be realized as ambiguity functions for arbitrary signals. In addition, a general synthesis method for the entire t, Ω-plane is also given.

Assume that the ambiguity function $\chi(t, \Omega)$ must approach a certain function $G(t, \Omega)$ given completely (both in magnitude and phase) over the entire t, Ω-plane. The integrated squared difference between these functions,

$$\varepsilon = \frac{1}{2\pi} \iint |G(t, \Omega) - \chi(t, \Omega)|^2 \, d\Omega \, dt, \tag{37–1}$$

will be considered as the approximation measure of the fit, so that the synthesis reduces to finding a signal $s(t)$ whose ambiguity function results in this integral being a minimum. We shall assume that the sought-after signal is formed by the superposition of the orthogonal functions

$$s(t) = \sum_k c_k f_k(t); \tag{37–2}$$

moreover,

$$\int_{-\infty}^{\infty} f_k(t) f_l^*(t) \, dt = \delta_{kl} = \begin{cases} 1 & \text{for} \quad k = l, \\ 0 & \text{for} \quad k \neq l \end{cases} \tag{37–3}$$

with

$$c_k = \int_{-\infty}^{\infty} s(t) f_k^*(t) \, dt.$$

Let us now determine the ambiguity function $\chi(t, \Omega)$ of the signal $s(t)$ represented in the form of (37–2). Disregarding the normalization condition (the $1/E$ factor), we can write

$$\begin{aligned}\chi(t, \Omega) &= \int s\left(t' + \frac{t}{2}\right) s^*\left(t' - \frac{t}{2}\right) e^{j\Omega t'}\, dt' \\ &= \sum_i c_i \sum_j c_j^* \int f_i\left(t' + \frac{t}{2}\right) f_j^*\left(t' - \frac{t}{2}\right) e^{j\Omega t'}\, dt' \\ &= \sum_{i,j} c_i c_j^* K_{ij}(t, \Omega), \end{aligned} \qquad (37\text{–}4)$$

where

$$K_{ij} = \int f_i\left(t' + \frac{t}{2}\right) f_j^*\left(t' - \frac{t}{2}\right) e^{j\Omega t'}\, dt'. \qquad (37\text{–}5)$$

Obviously, $K_{ij}(t, \Omega)$ is a cross product function of the two base signal functions, $f_i(t)$ and $f_j(t)$. Some of the properties of such functions were considered in §10. In particular, it was shown that for the two-dimensional Fourier transform of the product $K_{ij}(t, \Omega) \times K_{pq}^*(t, \Omega)$ we have the following equation [*cf.* (10–4)]:

$$\frac{1}{2\pi} \iint K_{ij}(t, \Omega)\, K_{pq}^*(t, \Omega)\, e^{j(\Omega t_1 - t\Omega_1)}\, dt\, d\Omega = K_{qj}(t_1, \Omega_1)\, K_{pi}^*(t_1, \Omega_1).$$

Setting $t_1 = \Omega_1 = 0$, we obtain

$$\frac{1}{2\pi} \iint K_{ij}(t, \Omega)\, K_{pq}^*(t, \Omega)\, dt\, d\Omega = K_{qj}(0, 0)\, K_{pi}^*(0, 0) = \int f_q f_j^*\, dt \int f_p^* f_i\, dt.$$

By virtue of the orthogonality condition for the f_k functions, we find

$$\frac{1}{2\pi} \iint K_{ij} K_{pq}^*\, dt\, d\Omega = \delta_{qj} \delta_{pi} = \begin{cases} 1 & \text{for} \quad i = p \quad \text{and} \quad j = q, \\ 0 & \text{for all other combinations of } i \text{ and } j. \end{cases} \qquad (37\text{–}6)$$

Thus, the family of functions in two variables, $\{K_{ij}(t, \Omega)\}$, is orthogonal. The functions $\{K_{ij}(t, \Omega)\}$ play an important role in the following. Referring to Sussman, we shall call them the *derivative base functions*; these functions are related to the system of signal functions $\{f_k(t)\}$.

From this representation, it is possible to formulate readily the necessary and sufficient condition for realizing the given function $G(t, \Omega)$ as the ambiguity function. The arbitrary function $G(t, \Omega)$ can be expanded in the base system $\{K_{ij}(t, \Omega)\}$, *i.e.*, represented in the form

$$G(t, \Omega) = \sum_{i,j} G_{ij} K_{ij}(t, \Omega).$$

The coefficients of this system form the matrix,

$$\mathbf{G} = \left\| \begin{matrix} G_{11} G_{12} G_{13} \ldots \\ G_{21} G_{22} G_{23} \ldots \\ \ldots\ldots\ldots\ldots \\ G_{i1}\ G_{i2}\ G_{i3}\ \ldots \\ \ldots\ldots\ldots\ldots \\ \ldots\ldots\ldots\ldots \end{matrix} \right\|$$

Eq. (37–4) shows that the function $G(t, \Omega)$ is the ambiguity function if and only if the matrix elements have the form

$$G_{ij} = c_i c_j^* . \tag{37–7}$$

In matrix terminology, this condition is stated as follows. *The function $G(t, \Omega)$ is realizable as the ambiguity function if the matrix* **G** *is a Hermitian matrix and is formed by the product of the column matrix*

$$\mathbf{c} = \left\| \begin{matrix} c_1 \\ c_2 \\ \cdot \\ \cdot \\ \cdot \end{matrix} \right\|$$

*and the transposed conjugate row matrix,**

$$\tilde{\mathbf{c}}^* = \| c_1^* c_2^* \ldots \| .$$

In addition, it is known that the rank of a matrix formed by such a multiplication is equal to unity; the converse is also valid: a matrix of unity rank can always be formed by multiplying two vectors. Therefore, $G(t, \Omega)$ *is a realizable ambiguity function if and only if* **G** *is a Hermitian matrix of unity rank.*

It is noted that by virtue of the orthogonality condition (37–6), the elements of the matrix G_{ij} are defined by

$$G_{ij} = \frac{1}{2\pi} \iint G(t, \Omega)\, K_{ij}^*(t, \Omega)\, dt\, d\Omega . \tag{37–8}$$

Of course, the realizability of a given function is an exceptional case. More frequently, $G(t, \Omega)$ cannot be realized exactly, and then there arises the

* It must be kept in mind that the matrix whose rows are the columns of the original matrix **G** is called the *transposed* matrix; the Hermitian matrix satisfies the condition $G_{ij} = G_{ji}^*$. The *rank* of the matrix is the lowest order of the non-zero determinant formed from this matrix by eliminating a certain number of rows and columns. For the matrix with unity rank, all determinants of the second order equal zero. Only the first order determinants, *i.e.*, the elements themselves of the matrix, differ from zero.

problem of minimizing the quadratic error (37–1). Considering (37–4), we can write

$$\varepsilon = \frac{1}{2\pi}\iint |G - \chi|^2 \, dt \, d\Omega$$
$$= \frac{1}{2\pi}\iint \Big(G - \sum_{ij} c_i c_j^* K_{ij}\Big)\Big(G^* - \sum_{p,q} c_p^* c_q K_{pq}^*\Big) dt \, d\Omega .$$

It is necessary to determine the coefficients c_k (or stated differently, the column matrix **c**) which brings this integral to a minimum. Assuming that c_k and c_k^* are independent variables,* the derivative of the error is

$$\frac{\partial \varepsilon}{\partial c_k^*} = \frac{1}{2\pi}\iint \Big[-\Big(G^* - \sum_{p,q} c_p^* c_q K_{pq}^*\Big)\sum_i c_i K_{ik}$$
$$- \Big(G - \sum_{i,j} c_i c_j^* K_{ij}\Big)\sum_q c_q K_{kq}^*\Big] \, dt \, d\Omega .$$

Using the orthogonality condition (37–6) and the notation of (37–8), we obtain, after some simple transformations,

$$\frac{\partial \varepsilon}{\partial c_k^*} = -\sum_i c_i (G_{ki} + G_{ik}^*) + 2c_k \sum_i |c_i|^2 .$$

Equating this derivative to zero and considering that $\sum_i |c_i|^2$ is the signal energy, E, we find the following condition for a maximum:

$$\tfrac{1}{2}\sum_i c_i (G_{ki} + G_{ik}^*) = E c_k . \qquad (37\text{–}9)$$

By investigating the derivative with respect to c_k in an analogous manner, an equivalent conjugate equation is obtained. The system of equations (37–9) (for $k=1, 2, 3, \ldots$) can be written in matrix form as follows:

$$\tfrac{1}{2}(\mathbf{G} + \tilde{\mathbf{G}}^*)\mathbf{c} = E\mathbf{c} . \qquad (37\text{–}10)$$

The solution of this system of homogeneous equations results in the sought-after vector **c**, *i.e.*, the expansion of the vector coefficients. The solutions are only possible for definite values of E, which are the eigenvalues of the matrix

$$\mathbf{G}' = \tfrac{1}{2}(\mathbf{G} + \tilde{\mathbf{G}}^*) . \qquad (37\text{–}11)$$

If **G** is a Hermitian matrix (*i.e.*, $\mathbf{G} = \tilde{\mathbf{G}}^*$), then obviously $\mathbf{G}' = \mathbf{G}$, and (37–11) takes the form

$$\mathbf{G}\mathbf{c} = E\mathbf{c} . \qquad (37\text{–}12)$$

* See footnote on p. 164.

The self-conjugacy of the matrix **G** has a simple meaning. Setting $G_{ik} = G_{ki}$, we obtain, on the basis of (37–5),

$$\begin{aligned} G^*(-t, -\Omega) &= \sum_{i,k} G_{ik}^* K_{ik}^*(-t, -\Omega) \\ &= \sum_{i,k} G_{ki} \int f_i^*\left(t' - \frac{t}{2}\right) f_k\left(t' + \frac{t}{2}\right) e^{j\Omega t'}\, dt' \\ &= \sum_{i,k} G_{ki} K_{ki}(t, \Omega) = G(t, \Omega). \end{aligned}$$

As mentioned in §10, this kind of symmetry is a necessary property of the ambiguity function. Consequently, the Hermitian requirement of matrix **G** is equivalent to the condition that the given function $G(t, \Omega)$ also have such symmetry. Of course, to obtain a better approximation, it is advantageous to use $G(t, \Omega)$ functions having such a property. However, this condition is not compulsory; the arbitrary function $G(t, \Omega)$ can be approximated by substituting for the non-Hermitian matrix **G** in (37–12), the Hermitian matrix **G**′ defined in (37–11). Equation (37–12) will now be considered, while keeping in mind the possibility of the above substitution.

The set of homogeneous equations (37–12) has solutions only when the determinant of the system is equal to zero, *i.e.*,

$$\begin{vmatrix} G_{11} - E & G_{12} & G_{13} & \cdots \\ G_{21} & G_{22} - E & G_{23} & \cdots \\ G_{31} & G_{32} & G_{33} - E & \cdots \\ \cdots & \cdots & \cdots & \cdots \\ \cdots & \cdots & \cdots & \cdots \end{vmatrix} = 0. \qquad (37\text{–}13)$$

This condition defines the admissible values of E. Since (37–13) is a higher order equation,* n eigenvalues $(E_1, E_2, \ldots, E_n)$ are obtained. The **G** matrix is Hermitian and, therefore, all the eigenvalues are real numbers. It can be readily shown that the solution corresponding to the maximum eigenvalue, $E_{\max}$, yields the best approximation of the desired function. In fact, by considering (37–8), it is possible to convert (37–1) into

$$\begin{aligned} \varepsilon = \frac{1}{2\pi} \iint |G(t, \Omega)|^2\, dt\, d\Omega \\ + \sum_{i,j} \sum_{p,q} c_i c_j^* c_p^* c_q \iint K_{ij}(t, \Omega)\, K_{pq}^*(t, \Omega)\, dt\, d\Omega \\ - \sum_{i,j} c_i^* c_j (G_{ij} + G_{ji}^*). \end{aligned}$$

* It is always possible to confine ourselves to a finite number of terms n in the initial expansion (37–2). Equation (37–13) has terms to the n–th power.

The orthogonality condition (37–6) and Eq. (37–9) lead to

$$\varepsilon = \frac{1}{2\pi}\iint |G(t, \Omega)|^2 \, dt \, d\Omega - E^2 .$$

Finally, without loss of generality, it can be assumed that the given function $G(t, \Omega)$ is normalized by the condition

$$\frac{1}{2\pi}\iint |G(t, \Omega)|^2 \, dt \, d\Omega = 1 ;$$

therefore,

$$\varepsilon = 1 - E^2 > 0 . \tag{37–14}$$

Consequently, the approximation of the given function will be optimum if the energy of the signal, E, is maximum.

Thus, the synthesis problem is reduced to determining the eigenvector **c** of the matrix **G** (or **G**′); this eigenvector corresponds to the maximum eigenvalue, $E_{\max}$. The magnitude of the error for the optimal case is defined by (37–14). If the order of the matrix is not high, then the roots of the determinant (37–13) can be found in the usual manner. However, it is noted that the methods for determining the eigenvalues and eigenvectors of matrices have been thoroughly developed; if necessary, solutions can be obtained even for comparatively complicated cases by the use of numerical methods (*e.g.*, see But [7], p. 115).

[§38] GENERALIZATION OF SYNTHESIS FROM THE MAGNITUDE OF AN AMBIGUITY FUNCTION

The method for synthesizing signals presented in the previous section has one essential disadvantage. It assumes that the function $G(t, \Omega)$ is given completely, *i.e.*, both in magnitude and phase. It has been noted repeatedly that the phase of the ambiguity function is of no significance and, consequently, in constructing the approximation there cannot be any certainty that the solution is optimum, since for another phase distribution of $G(t, \Omega)$ [and respectively $\chi(t, \Omega)$], a better approximation is possible with respect to the amplitude distribution. A direct synthesis method formulated from the magnitude $|G(t, \Omega)|$ does not lead to any satisfactory results. However, a procedure can be proposed which improves the approximation at the expense of the phase distribution. In order to proceed with this problem, it is desirable to investigate in more detail the synthesis method considered in §37. Instead of using the initial base system functions of the signal, $f_k(t)$

[which will henceforth be denoted by $f_k^{(0)}(t)$], it is possible to form a new system of base functions $f_k^{(1)}(t)$ with the aid of the linear transformation

$$f_k^{(1)}(t) = \sum_p Q_{pk} f_p^{(0)}(t). \tag{38–1}$$

For the time being, no conditions are imposed on the coefficients of this transformation Q_{pk}. Introducing the matrix

$$\mathbf{Q} = \begin{Vmatrix} Q_{11} & Q_{12} & Q_{13} \cdots \\ Q_{21} & Q_{22} & Q_{23} \cdots \\ Q_{31} & Q_{32} & Q_{33} \cdots \\ \cdots & \cdots & \cdots \\ \cdots & \cdots & \cdots \end{Vmatrix},$$

the transformation (38–1) can be represented in matrix form which establishes the relationship between the vectors $\mathbf{f}^{(0)}(t)$ and $\mathbf{f}^{(1)}(t)$:

$$\mathbf{f}^{(1)}(t) = \tilde{\mathbf{Q}}\mathbf{f}^{(0)}(t). \tag{38–2}$$

It is also required that the system of functions $\{f^{(1)}(t)\}$ be orthonormal, *i.e.*,

$$\int f_k^{(1)}(t)\, f_l^{(1)*}(t)\, dt = \delta_{kl} = \begin{cases} 1 & \text{for} \quad k = l, \\ 0 & \text{for} \quad k \neq l. \end{cases} \tag{38–3}$$

Taking the orthogonality of the initial system $\{f_k^{(0)}(t)\}$ into account, we obtain

$$\int f_k^{(1)}(t)\, f_l^{(1)*}(t)\, dt = \sum_p \sum_q Q_{pk} Q_{ql}^* \int f_p^{(0)}(t)\, f_q^{(0)*}(t)\, dt$$
$$= \sum_p Q_{pk} Q_{pl}^*.$$

Therefore, (38–3) leads to

$$\sum_p Q_{pk} Q_{pl}^* = \delta_{kl} = \begin{cases} 1 & \text{for} \quad k = l, \\ 0 & \text{for} \quad k \neq l, \end{cases}$$

which in matrix form can be written as follows:

$$\tilde{\mathbf{Q}}\mathbf{Q}^* = \begin{Vmatrix} 1 & 0 & 0 \ldots \\ 0 & 1 & 0 \ldots \\ 0 & 0 & 1 \ldots \\ \cdots & \cdots & \cdots \\ \cdots & \cdots & \cdots \end{Vmatrix}. \tag{38–4}$$

The above is called the unitary transformation and can be given a simple geometrical interpretation. The initial system of orthonormal base functions $\{f_k^{(0)}(t)\}$ can be made similar to the system of orthogonal coordinate vectors of unit amplitude in the multi-dimensional space. The expansion of the signal $s(t)$ in the $\{f_k^{(0)}(t)\}$ system corresponds to the expansion of a vector in a system of coordinate vectors. The c_k coefficients in (37–2) are the projections of the vector **c** onto the axes. The new system $\{f_k^{(1)}(t)\}$ obtained by the unitary transformation from $\{f_k^{(0)}(t)\}$ is also similar to the system of orthogonal unit vectors, and every vector (signal) can be expanded in this new system. Consequently, the unitary transformation $\mathbf{Q}$, converting $\{f_k^{(0)}\}$ into $\{f_k^{(1)}\}$, is equivalent to the rotation of the system coordinate vectors in the multi-dimensional space. Moreover, the orthogonality and the unit amplitude of the vectors are preserved. A set of all unitary transformations forms a group. This means that any two consecutive transformations (rotations) $\mathbf{Q}_1$ and $\mathbf{Q}_2$ are equivalent to the one transformation (rotation), *i.e.*, $\mathbf{Q}_3 = \mathbf{Q}_1 \cdot \mathbf{Q}_2$; in addition, for any transformation $\mathbf{Q}$, there exists an inverse transformation $\mathbf{Q}^{-1}$ converting the $\{f_k^{(1)}(t)\}$ system into the initial $\{f_k^{(0)}(t)\}$ system. It follows from (38–4) that the matrix of the inverse unitary transformation coincides with the adjoint matrix:

$$\mathbf{Q}^{-1} = \tilde{\mathbf{Q}}^* .$$

The signal's new base system, $\{f_k^{(1)}(t)\}$, generates another new system of base functions $\{K_{ij}^{(1)}(t, \Omega)\}$ given by the form

$$K_{ij}^{(1)}(t, \Omega) = \int f_i^{(1)}\left(t' + \frac{t}{2}\right) f_j^{(1)*}\left(t' - \frac{t}{2}\right) \mathrm{e}^{j\Omega t'}\, dt'$$

or using (38–1),

$$K_{ij}^{(1)}(t, \Omega) = \sum_p \sum_q Q_{pi} Q_{qj}^* K_{pq}^{(0)}(t, \Omega) .$$

It is readily seen that in matrix notation this is equivalent to

$$\mathbf{K}^{(1)}(t, \Omega) = \tilde{\mathbf{Q}} \mathbf{K}^{(0)}(t, \Omega)\, \mathbf{Q}^* . \qquad (38\text{–}5)$$

Multiplying the left- and right-hand sides of this equation by $\mathbf{Q}^*$ and $\tilde{\mathbf{Q}}$, respectively, and taking into account (38–4), we find the inverse transformation from $\mathbf{K}^{(1)}$ to $\mathbf{K}^{(0)}$:

$$\mathbf{K}^{(0)}(t, \Omega) = \mathbf{Q}^* \mathbf{K}^{(1)}(t, \Omega)\, \tilde{\mathbf{Q}} . \qquad (38\text{–}6)$$

The given function $G(t, \Omega)$ can be expanded in the new system of derivative base functions $\{K_{ij}^{(1)}(t, \Omega)\}$. To find the coefficients of this expansion, $G_{ij}^{(1)}$, it is necessary to use matrix calculus. From (37–8), the matrices of the coefficients $G^{(0)}$ and $G^{(1)}$, pertaining to the expansion in the $\{K_{ij}^{(0)}\}$ and $\{K_{ij}^{(1)}\}$ systems, can be written in the form of integrals which contain the matrices of $\mathbf{K}^{(0)}$ and $\mathbf{K}^{(1)}$:

$$\mathbf{G}^{(0)} = \frac{1}{2\pi} \iint G(t, \Omega)\, \mathbf{K}^{(0)*}(t, \Omega)\, dt\, d\Omega ,$$

$$\mathbf{G}^{(1)} = \frac{1}{2\pi} \iint G(t, \Omega)\, \mathbf{K}^{(1)*}(t, \Omega)\, dt\, d\Omega .$$

Using (38–5), we find

$$\mathbf{G}^{(1)} = \frac{1}{2\pi}\iint G(t, \Omega)\,\tilde{\mathbf{Q}}^*\mathbf{K}^{(0)*}(t, \Omega)\,\mathbf{Q}\;dt\;d\Omega.$$

Since the matrices $\tilde{\mathbf{Q}}^*$ and $\mathbf{Q}$ do not depend on t and Ω, they can be taken outside the integral sign. As a result, we obtain

$$\mathbf{G}^{(1)} = \tilde{\mathbf{Q}}^*\mathbf{G}^{(0)}\mathbf{Q}. \tag{38–7}$$

This matrix expression establishes the relationship between the coefficients $G_{ij}^{(1)}$ and $G_{ij}^{(0)}$. In the expanded form, such a dependence is expressed by

$$G_{ij}^{(1)} = \sum_p \sum_q Q_{pi}^* Q_{qj} G_{pq}^{(0)}.$$

The inverse transformation to (38–7), which converts $\mathbf{G}^{(1)}$ into $\mathbf{G}^{(0)}$, is

$$\mathbf{G}^{(0)} = \mathbf{Q}\mathbf{G}^{(1)}\tilde{\mathbf{Q}}^*. \tag{38–8}$$

The subsequent discussion is based on the diagonalization of matrices (*e.g.*, see Courant & Hilbert [36], p. 11). For any matrix $\mathbf{G}^{(0)}$, it is possible to choose a unitary transformation $\mathbf{Q}^{(0)}$ for which the transformed matrix $\mathbf{G}^{(1)}$ takes the diagonal form

$$\mathbf{G}^{(1)} = \tilde{\mathbf{Q}}^{(0)*}\mathbf{G}^{(0)}\mathbf{Q}^{(0)} = \mathbf{\Lambda} = \begin{Vmatrix} E_1 & 0 & 0 & \ldots \\ 0 & E_2 & 0 & \ldots \\ 0 & 0 & E_3 & \ldots \\ \ldots & \ldots & \ldots & \ldots \\ \ldots & \ldots & \ldots & \ldots \end{Vmatrix}. \tag{38–9}$$

The matrix $\mathbf{Q}^{(0)}$ is called the *diagonalizing operator* for the matrix $\mathbf{G}^{(0)}$, and a direct method is known for determining $\mathbf{Q}^{(0)}$ in terms of $\mathbf{G}^{(0)}$. All elements of the matrix $\mathbf{\Lambda}$ are equal to zero except the principal diagonal elements which are equal to the eigenvalues E_ν of the initial matrix $\mathbf{G}^{(0)}$. Now it is possible to formulate a new process for approximating the given function $G(t, \Omega)$.

Let us require that the magnitude $|\chi(t, \Omega)|$ approaches the given amplitude function $|G(t, \Omega)|$ in the best possible way. To begin with, let us assign to the magnitude $|G(t, \Omega)|$ a certain arbitrary phase $\psi_0(t, \Omega)$ to form the function

$$G^{(0)}(t, \Omega) = |G(t, \Omega)|\, e^{j\psi_0(t, \Omega)}.$$

By expanding the above in the initial system of derivative base functions, $\{K_{ij}^{(0)}(t, \Omega)\}$, which is related to $\{f_k^{(0)}(t)\}$, we determine the matrix of the coefficients of $\mathbf{G}^{(0)}$. As we have seen, the best approximation is yielded by the eigenvector of the matrix $\mathbf{G}^{(0)}$, corresponding to the maximum energy value, $E_{\max}$.

The matrix $\mathbf{G}^{(0)}$ is diagonalized by using the corresponding unitary transformation $\mathbf{Q}^{(0)}$, *i.e.*, we form the matrix

$$\mathbf{G}^{(1)} = \tilde{\mathbf{Q}}^{(0)*}\mathbf{G}^{(0)}\mathbf{Q}^{(0)} = \left\| \begin{matrix} E_1^{(0)} & 0 & 0 \ldots \\ 0 & E_2^{(0)} & 0 \ldots \\ 0 & 0 & E_3^{(0)} \ldots \\ \ldots\ldots & \ldots\ldots & \ldots\ldots \\ \ldots\ldots & \ldots\ldots & \ldots\ldots \end{matrix} \right\| .$$

This unitary transformation generates simultaneously a new system of base functions in the vector space of the signals,

$$\mathbf{f}^{(1)}(t) = \tilde{\mathbf{Q}}^{(0)}\mathbf{f}^{(0)}(t)$$

and a new system of derivative base functions, *i.e.*,

$$\mathbf{K}^{(1)}(t, \Omega) = \tilde{\mathbf{Q}}^{(0)}\mathbf{K}^{(0)}(t, \Omega)\,\mathbf{Q}^{(0)*} .$$

Since $\mathbf{G}^{(1)}$ is a diagonal matrix, the expansion of the function $G^{(0)}(t, \Omega)$ in $\{K_{ij}^{(1)}(t, \Omega)\}$ contains only diagonal terms:

$$G^{(0)}(t, \Omega) = \sum_{\nu} E_{\nu}^{(0)} K_{\nu\nu}^{(1)}(t, \Omega) . \tag{38–10}$$

Consequently, in order to find the best approximation, it is now sufficient to choose the term of expansion (38–10) which has the largest coefficient, $E_{\max}$ (the largest eigenvalue). As a matter of fact, this term is the ambiguity function $\chi^{(0)}(t, \Omega)$, which provides the best approximation of $G^{(0)}(t, \Omega)$:

$$\chi^{(0)}(t, \Omega) = (E_{\nu}^{(0)})_{\max} K_{\nu\nu}^{(1)}(t, \Omega) .$$

The corresponding optimal signal is one of the base vectors of the $f^{(1)}$ system:

$$s^{(0)}(t) = f_{\nu}^{(1)}(t) = \sum_{p} Q_{p\nu}^{(0)} f_p^{(0)}(t) .$$

It follows from (38–10) that the eigenvalue $E_{\nu}^{(0)}$ is given in the form

$$E_{\nu}^{(0)} = \frac{1}{2\pi} \iint G^{(0)}(t, \Omega)\, K_{\nu\nu}^{(1)*}(t, \Omega)\, dt\, d\Omega . \tag{38–11}$$

Let us now attempt to modify the phase of the given function $G^{(0)}(t, \Omega)$ in such a manner as to increase the maximum eigenvalue E_{ν} as much as possible. Since $|G(t, \Omega)|$ and $K_{\nu\nu}^{(1)}(t, \Omega)$ are given, it is obvious that the integral (38–11) is maximum if the function

$$G^{(0)}(t, \Omega) = |G(t, \Omega)|\, e^{j\psi_0(t, \Omega)}$$

is replaced by

$$G^{(1)}(t, \Omega) = |G(t, \Omega)|\, e^{j\psi_1(t, \Omega)} \tag{38–12}$$

whose phase $\psi_1(t, \Omega)$ coincides with that of the base function

$$K_{\nu\nu}^{(1)}(t, \Omega) = |K_{\nu\nu}^{(1)}(t, \Omega)| \, e^{j\psi_1(t, \Omega)} .$$

Moreover, the integrand of (38–11) is replaced by the magnitude $|G^{(0)}K_{\nu\nu}^{(1)}|$ which maximizes the integral.

Thus, it is advantageous to replace the initial function $G^{(0)}(t, \Omega)$ by the new function $G^{(1)}(t, \Omega)$ which has, from (38–12), the same magnitude but a different phase. The following approximation process is similar to the previous procedure, but now $\{K_{ij}^{(1)}(t, \Omega)\}$ must be used as the initial base system.

By expanding $G^{(1)}(t, \Omega)$ in this system, we determine the matrix $\mathbf{G}^{(1)}$. The corresponding unitary transformation $\mathbf{Q}^{(1)}$ reduces this matrix to the diagonal form,

$$\mathbf{G}^{(2)} = \tilde{\mathbf{Q}}^{(1)*}\mathbf{G}^{(1)}\mathbf{Q}^{(1)} = \left\| \begin{matrix} E_1^{(1)} & 0 & 0 & \ldots \\ 0 & E_2^{(1)} & 0 & \ldots \\ 0 & 0 & E_3^{(1)} & \ldots \\ \ldots & \ldots & \ldots & \ldots \\ \ldots & \ldots & \ldots & \ldots \end{matrix} \right\| ,$$

and generates a new base system $\mathbf{f}^{(2)}(t)$ as well as a new derivative base system $\mathbf{K}^{(2)}(t, \Omega)$. In this system, the expansion $G^{(1)}(t, \Omega)$ contains only the diagonal terms

$$G^{(1)}(t, \Omega) = \sum_{\nu} E_\nu^{(1)} K_{\nu\nu}^{(2)}(t, \Omega)$$

and, as the next approximation, we must take the base function $K_{\nu\nu}^{(1)}$ having the maximum eigenvalue $(E_\nu^{(2)})_{\max}$:

$$\chi^{(2)}(t, \Omega) = (E_\nu^{(1)})_{\max} K_{\nu\nu}^{(2)}(t, \Omega) .$$

Furthermore, one can proceed to the next step of the approximation by replacing the function $G^{(1)}(t, \Omega)$ by

$$G_2^{(2)}(t, \Omega) = |G(t, \Omega)| \, e^{j\psi_2(t, \Omega)} ,$$

where $\psi_2(t, \Omega)$ is the phase of the base function $K_{\nu\nu}^{(2)}(t, \Omega)$, etc.

It was shown by Sussman [56] that such an iteration process of approximation converges, *i.e.*, it yields an increasing sequence of eigenvalues $(E_\nu^{(0)})_{\max}, (E_\nu^{(1)})_{\max}, (E_\nu^{(2)})_{\max}, \ldots$ approaching a certain limit. This means that after several iteration steps, the eigenvalue growth slows down to the extent that further approximations are meaningless. Therefore, by confining ourselves to n iterations, we can return to the initial base system of the signal by sequencing m unitary transformations, *i.e.*,

$$\mathbf{f}^{(m)}(t) = \tilde{\mathbf{Q}}^{(m)} \ldots \tilde{\mathbf{Q}}^{(1)}\tilde{\mathbf{Q}}^{(0)}\mathbf{f}^{(0)}(t)$$

and, by the same token, express the determined signal in terms of the initial base functions [since $s^{(m)}(t)=f_\nu^{(m)}(t)$]. From (37–14), the error of such an approximation is

$$\varepsilon = 1 - (E_\nu^{(m)})^2_{\max}. \qquad (38\text{–}13)$$

Since the maximum eigenvalue increases by sequencing, the m–th iteration yields a better approximation than the zero iteration. However, it should not be assumed that such a process makes it possible to decrease arbitrarily the approximation error. On the contrary, the limit of the error ε tends to depend on both the given magnitude $|G(t, \Omega)|$ and on the phase $\psi_0(t, \Omega)$ which was initially assigned to this magnitude. If the error obtained from the iteration is too large, then a new series of iterations must be attempted proceeding from another initial phase function.

Thus, if the synthesis is carried out for the function $G(t, \Omega)$ which is completely given (both in magnitude and phase), and if this method realizes the best (but not necessarily a satisfactory!) quadratic approximation, then the method of successive approximations must be used where only the magnitude $|G(t, \Omega)|$ is given; moreover, every approximation series does not guarantee the necessary upper limit of the error. In spite of this deficiency (associated with the non-realizability of given functions), the above method is the most complete of the known solutions for synthesizing signals from both complete functions, $G(t, \Omega)$, and function magnitudes, $|G(t, \Omega)|$. Other methods are inferior to the one above as far as generality is concerned; they solve special problems and are confined to a relatively narrow class of signals.

It is also noted that the method considered requires cumbersome calculations. It becomes necessary to solve systems of equations with a large number of unknowns, to evaluate double integrals, etc. Therefore, in the majority of cases, it is expedient to use modern computer techniques. In this connection, a discrete formulation of the problem, which simplifies the computer programming for solving the synthesis, is given by Sussman [56].

Appendices

Appendix 1

Spectra of Wide-Band FM Signals Approximated by the Stationary Phase Method

Let us consider the integral

$$I = \int_{-\infty}^{\infty} a(t)\, e^{jp\varphi(t)}\, dt\,, \tag{A1–1}$$

where p is a very large number ($p \gg 1$), and $a(t)$ and $\varphi(t)$ are comparatively slowly varying functions. The function

$$e^{jp\varphi(t)} = \cos p\varphi(t) + j \sin p\varphi(t)$$

is an oscillatory process with the frequency being directly proportional to p. For large values of p, the net effect of the oscillation over a complete cycle is negligibly small (effectively, the positive and negative portion of successive half cycles cancel each other). However, this compensation of adjacent half waves is not valid near the *points of stationary phase* defined by

$$\varphi'(t_0) = 0\,. \tag{A1–2}$$

The "instantaneous frequency" at these points equals zero and the oscillation process stops (more accurately, the function ceases to oscillate). The greater the value of p, the more reliable the assumption that the value of the integral (A1–1) is defined only by the contributions in the neighborhood of the stationary phase points which satisfy the condition (A1–2).

Near the stationary phase point t_0, the function $\varphi(t)$ can be represented by the first few terms of its Taylor expansion:

$$\varphi(t) \approx \varphi(t_0) + \frac{\varphi''(t_0)}{2}(t - t_0)^2\,.$$

The linear term is absent by virtue of condition (A1–2). Therefore, the contribution made to integral (A1–1) in the Δ neighborhood of the stationary phase point t_0 takes the following form:

$$a(t_0) \int_{t_0-\frac{\Delta}{2}}^{t_0+\frac{\Delta}{2}} e^{jp\left[\varphi(t_0)+\frac{\varphi''(t_0)}{2}(t-t_0)^2\right]} dt; \tag{A1–3}$$

moreover, the factor $a(t)$ is assumed to be constant for the integration and thus is removed from under the integral sign. Also the integration limits of (A1–3) may be arbitrarily expanded without substantially changing the value of the integral. This is due to the fact that when removed from the vicinity of the point t_0, the integrand of (A1–3) again becomes a rapidly oscillating function with the adjacent half waves cancelling each other. Thus, if (A1–2) has only one solution, then the following is the approximation obtained for the integral (A1–1):

$$I \approx a(t_0)\, e^{jp\varphi(t_0)} \int_{-\infty}^{\infty} e^{jp\frac{\varphi''(t_0)}{2}(t-t_0)^2} dt$$

$$= a(t_0)\sqrt{\frac{2\pi}{|p\varphi''(t_0)|}}\, e^{j\left[p\varphi(t_0)\pm\frac{\pi}{4}\right]}. \tag{A1–4}$$

Here the plus sign applies when $p\varphi''(t_0)>0$, and the minus sign when $p\varphi''(t_0)<0$.

If (A1–2) has several solutions (*i.e.*, there are several stationary phase points), then it is necessary to take the sum of terms similar to (A1–4).

It is not possible to discuss here the limitations of the stationary phase method. However, it should be noted that the main requirement for using this method is that $a(t)$ be slowly varying in the vicinity of the stationary phase point. This condition is expressed mathematically by the following:

$$|p\varphi''(t_0)| \gg \left(\frac{a'(t_0)}{a(t_0)}\right)^2. \tag{A1–5}$$

It is clear from the above that the stationary phase method can also be applied to integrals with finite limits, provided the stationary phase points lie within the area of integration and that these points are not too close to the boundaries of this area. It is also noted that the parameter p in (A1–1) to (A1–5) does not have to be in explicit form. The product $p\varphi(t)$ can be replaced by a new function $\varphi_1(t)$. It is important, however, that the exponential factor in the integral (A1–1) be oscillating rapidly compared to the amplitude factor.

The stationary phase method is now applied to determine the spectra of wide-band FM signals:

$$s(t) = a(t)\, e^{j\varphi_0(t)},$$

where $a(t)$ and $\varphi_0(t)$ are the amplitude and phase modulation functions, respectively. The spectrum of this signal is given by

$$S(\omega) = \int_{-\infty}^{\infty} s(t)\, e^{-j\omega t}\, dt = \int_{-\infty}^{\infty} a(t)\, e^{j[\varphi_0(t) - \omega t]}\, dt. \qquad \text{(A1–6)}$$

In this case the equation of the stationary phase points has the form

$$\varphi'(t_0) = \varphi_0'(t_0) - \omega = 0,$$

or

$$\varphi_0'(t_0) = \omega_c(t_0) = \omega. \qquad \text{(A1–7)}$$

Consequently, from the stationary phase method, the spectral density at the frequency ω is basically defined by the contributions during the times when the instantaneous frequency runs through the value ω. At other times, the instantaneous frequency is different from the selected frequency of the spectrum, and the signal form is not important. This situation is, apparently, in agreement with the qualitative "quasi-harmonic" interpretation of FM signals which considers the oscillations as having a slowly changing frequency. If the instantaneous frequency changes monotonically, then there is a single stationary point and (A1–4) becomes

$$S(\omega) = a(t_0) \sqrt{\frac{2\pi}{|\omega_c'(t_0)|}}\; e^{j\left[\varphi_0(t_0) - \omega t_0 \pm \frac{\pi}{4}\right]}. \qquad \text{(A1–8)}$$

This result can readily be confirmed by measurement.

For example, let the spectral analysis be carried out with the aid of a narrow-band resonant circuit. Then tune the circuit to resonate at the chosen frequency ω, and fix the instant of time t_0 when the instantaneous frequency coincides with ω. The energy arriving at the output of the resonant circuit is proportional to the square of the signal amplitude at the time t_0 and to the length of time the signal is in the circuit's pass band. The latter (time) is inversely proportional to the frequency rate of change at the time t_0. Therefore, neglecting constant coefficients, we have

$$|S(\omega)|^2 = \frac{a^2(t_0)}{|\omega_c'(t_p)|},$$

which agrees with (A1–8). In addition to the amplitude (energy) relationships, the stationary phase method also provides the phase structure of the spectrum. With this method (the simplest version is considered here), it is possible to calculate only the spectrum over the region which the signal's instantaneous frequency runs through. The energy density outside this region is low but, strictly speaking, is different from zero. However, the points of stationary phase are absent from this frequency region and, in the approximation used, the spectral function becomes zero.

Such an assumption is acceptable for wide-band FM pulses. For example, consider rectangular pulses with linear frequency modulation. In this case,

$$S(\omega) = \int_{-T/2}^{T/2} e^{[\gamma t^2 - (\omega - \omega_0)t]} \, dt. \tag{A1–9}$$

The point of stationary phase, t_0, is determined from the condition

$$\omega_c(t_0) = \omega_0 + 2\gamma t_0 = \omega$$

or

$$t_0 = \frac{\omega - \omega_0}{2\gamma};$$

from (A1–8), an approximate value of the spectrum is (γ is assumed to be positive)

$$S(\omega) \approx \begin{cases} \sqrt{\dfrac{\pi}{\gamma}}\, e^{j\left[\frac{(\omega-\omega_0)^2}{4\gamma} + \frac{\pi}{4}\right]} & \text{for} \quad \omega_0 - \gamma T < \omega < \omega_0 + \gamma T. \\ 0 & \text{for} \quad |\omega - \omega_0| > \gamma T. \end{cases}$$

Thus, for the approximation used, the amplitude spectrum has a rectangular form. The upper drawing in Figure 46 shows this spectrum (cross-hatched); the lower drawing is a plot of the signal's instantaneous frequency function. The relationship between the spectra and the signal form, which follows from the stationary phase method, is clearly seen in Figure 46. The constant level of the spectral density is due to the constant modulation rate and amplitude of the pulse.

The exact expression of this spectrum can also be derived. In fact, supplying the exponent in (A1–9) to complete the square, we obtain, after a simple transformation,

$$S(\omega) = \sqrt{\frac{\pi}{\gamma}}\, e^{-j\frac{(\omega-\omega_0)}{4\gamma}} [F(z_1) - F(z_2)].$$

Here

$$F(z) = \frac{2}{\sqrt{\pi}} \int_0^z e^{jx^2} \, dx$$

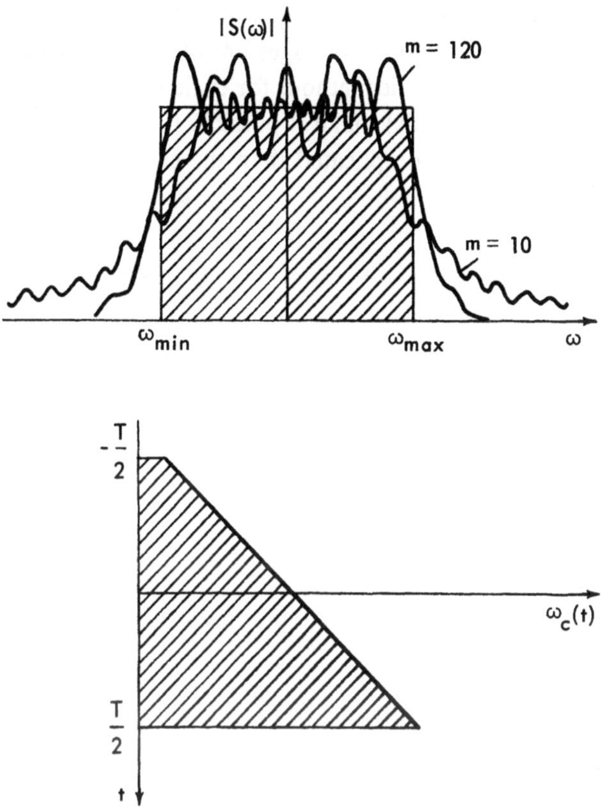

Fig. 46

is the Fresnel integral, and the arguments z_1 and z_2 are given by

$$z_1 = \frac{1}{2\sqrt{\gamma}}[\gamma T - (\omega - \omega_0)] = \frac{\sqrt{\pi m}}{2}\left(1 - 2\,\frac{\omega - \omega_0}{\Omega}\right),$$

$$z_2 = \frac{1}{2\sqrt{\gamma}}[-\gamma T - (\omega - \omega_0)] = -\frac{\sqrt{\pi m}}{2}\left(1 + 2\,\frac{\omega - \omega_0}{\Omega}\right),$$

where $\Omega = 2\gamma T$ is the total change of the instantaneous frequency over the pulse duration and $m = \Omega T/2\pi = \gamma T^2/\pi$ is the compression ratio.

Figure 46 also shows the plots of the exact spectra for two values of the compression ratio ($m = 10$ and $m = 120$) drawn from the above expressions (the graphs are taken from [34]). It can be seen from this figure that the results of the approximate and exact expressions are in good agreement for most important cases of large compression. The largest errors of the approximation occur at the frequencies near the beginning and end of the spectrum, *i.e.*, the region where the spectrum envelope undergoes sudden changes. The limitation of the approximate expression is due mainly to the bandwidth of the signal spectrum on which the frequency modulation exerts a prevailing effect; the amplitude changes are of secondary significance. It should be noted that asymptotic methods for evaluating integrals (stationary phase is one of these methods) make it possible to calculate the spectra of FM signals that are more complicated than the cases considered here (see [9]).

Appendix 2

Calculation of Phase Measurement Errors

Assume that the instantaneous frequency of an FM oscillator follows $f(t)$; it is necessary to measure this function with an appropriate device. No assumptions are made concerning $f(t)$, except that it is a continuous, monotonic function, and that the measurements are made during a certain time T (*e.g.*, over the duration of the FM pulse).

The measurement reduces to determining the instantaneous frequency $f(t)$ for various instants of time t in the time interval $0<t<T$. Moreover, time errors, x, and frequency errors, y, are inevitable. As a result, instead of $f(t)$, the following is measured:

$$f_1(t) = f(t + x) + y . \tag{A2–1}$$

Neither the measurement technique nor the distribution of the random errors x and y will be discussed here; it is assumed that such a distribution does exist and, consequently, that the expectation, variance, and other properties of the distribution can be found. In particular, it is assumed that the systematic errors of the measuring equipment have been eliminated so that the means of the errors are zero:

$$\bar{x} = \bar{y} = 0 . \tag{A2–2}$$

The interest here is in the phase error accumulated while making measurements over the time period T. Assuming no measurement errors, the net phase for this time period is

$$\psi(T) = \int_0^T f(t)\, dt .$$

Taking the measurement errors into account, the net phase measured for the same period of time is

$$\psi_1(T) = \int_0^T [f(t + x) + y]\, dt .$$

The error of the measurement, $\delta\psi=\psi_1-\psi_2$, is determined from

$$\delta\psi = \int_0^T [f(t+x) - f(t)]\, dt + \int_0^T y\, dt .$$

Considering that the error x is small, it can be assumed that

$$f(t+x) \approx f(t) + x f'(t)$$

and, therefore,

$$\delta\psi = \int_0^T x f'(t)\, dt + \int_0^T y\, dt = \delta\psi_1 + \delta\psi_2 . \qquad \text{(A2–3)}$$

It is noted that x and y are random functions of time and, consequently, cannot be taken out from under the integral sign.

Thus, the phase error consists of two terms: one depends only on the time measurement, and the other only on the frequency measurement. These terms are statistically independent since the frequency and time measurements are not related to each other (conducted as separate operations). The mean of the errors, $\overline{\delta\psi_1}$ and $\overline{\delta\psi_2}$, are equal to zero by virtue of (A2–2). Therefore, the mean square of the phase error is

$$\sigma_\psi^2 = \overline{\delta\psi^2} = \overline{(\delta\psi_1 + \delta\psi_2)^2} = \overline{\delta\psi_1^2} + \overline{2\delta\psi_1\delta\psi_2} + \overline{\delta\psi_2^2} = \overline{\delta\psi_1^2} + \overline{\delta\psi_2^2} . \qquad \text{(A2–4)}$$

Furthermore, it is found that

$$\overline{\delta\psi_1^2} = \int_{-\infty}^{\infty} \delta\psi_1^2 p(x)\, dx = \int_{-\infty}^{\infty} p(x) \left[\int_0^T x f'(t)\, dt\right]^2 dx ,$$

where $p(x)$ is the probability distribution of the error x.

Substituting a product of integrals for the square of the integrals and changing the order of integration, we find

$$\overline{\delta\psi_1^2} = \int_0^T f'(t_1)\, dt_1 \int_0^T f'(t_2)\, dt_2 \int_{-\infty}^{\infty} x^2 p(x)\, dx .$$

The last integral is the mean square of the time measurement error:

$$\int_{-\infty}^{\infty} x^2 p(x)\, yx = \overline{x^2} = \sigma_t^2 .$$

Therefore,

$$\overline{\delta\psi_1^2} = \sigma_t^2 \left[\int_0^T f'(t)\, dt \right]^2 = \sigma_t^2 [f(T) - f(0)]^2 = \sigma_t^2 \Omega^2 ,$$

where Ω is the frequency deviation during the time interval T.

Similarly, it can be shown that

$$\overline{\delta\psi_2^2} = \sigma_\omega^2 T^2 .$$

Finally, substituting the above in (A2–4) yields

$$\sigma_\psi^2 = \sigma_t^2 \Omega^2 + \sigma_\omega^2 T^2 .$$

This relationship is used in §16.

Appendix 3

Estimate of Amplitude Distribution for a Signal with Random Initial Phase

The expression

$$\hat{A}_0 = A_0\, e^{j\theta} + n_g\,, \tag{A3–1}$$

for evaluating the complex amplitude of a signal, was obtained in §20. The complex noise component n_g can be expressed by

$$n_g = C\, e^{j\gamma}\,.$$

The random amplitude of the noise C follows the Rayleigh distribution, and the phase γ has a uniform distribution. Substituting the above in (A3–1), we obtain

$$\hat{A}\, e^{-j\theta} = A_0 + n_g\, e^{-j\theta} = A_0 + C\, e^{j(\gamma-\theta)}\,. \tag{A3–2}$$

The difference between the random variables, $\gamma_1 = (\gamma - \theta)$, also has a uniform distribution. For example, this can readily be shown by combining two uniform distributions. Therefore, the noise component of (A3–2) has the same probability distribution as n_g. In other words, if the noise is written in the form of the complex number

$$C\, e^{j(\gamma-\theta)} = \xi + j\eta\,,$$

then each component has a normal distribution with identical variance σ^2 [see (18–9)]. These components are not correlated (by virtue of being in quadrature) and, consequently, are independent.

Furthermore, it follows from (A3–2) that

$$|\hat{A}|^2 = \hat{A}_0^2 = z = (A_0 + \xi)^2 + \eta^2 = z_1 + z_2\,.$$

The distributions of the random variables z_1 and z_2 can readily be determined from the known distributions of ξ and η (e.g., see Cramér [35], pp. 166, 233). If $z_1, z_2 > 0$, then

$$p_1(z_1) = \frac{1}{\sigma\sqrt{2\pi z_1}}\, e^{-\frac{z_1 + A_0^2}{2\sigma^2}} \cosh\frac{A_0\sqrt{z_1}}{\sigma^2}\,,$$

$$p_2(z_2) = \frac{1}{\sigma\sqrt{2\pi z_2}}\, e^{-\frac{z_2}{2\sigma^2}}\,.$$

The distribution of $z=(z_1+z_2)$ is expressed by the convolution

$$p(z)=\int_{-\infty}^{\infty} p_1(z_1)\, p_2(z-z_1)\, dz_1 .$$

After a few transformations, this becomes

$$p(z)=\frac{1}{\pi\sigma^2}\, e^{-\frac{A_0{}^2+z}{2\sigma^2}} \int_0^1 \cosh\left(\frac{A_0\sqrt{z}}{\sigma^2}\, x\right) \frac{dx}{\sqrt{1-x^2}}$$

$$=\frac{1}{2\sigma^2}\, e^{-\frac{A_0{}^2+z}{2\sigma^2}}\, I_0\left(\frac{A_0\sqrt{z}}{\sigma^2}\right).$$

Here the substitution for the value of the known integral is made (Gradshtein & Ryzhik [17], p. 355, §3.534, No. 2). By letting $\hat{A}_0=\sqrt{z}$, we finally obtain

$$p(\hat{A}_0)=\frac{\hat{A}_0}{\sigma^2}\, e^{-\frac{A_0{}^2+\hat{A}_0{}^2}{2\sigma^2}}\, I_0\left(\frac{A_0\hat{A}_0}{\sigma^2}\right).$$

Appendix 4

Determination of the Dolph-Chebyshev Autocorrelation Function and the Corresponding Power Spectra

From (27–8) and (27–9), the optimal autocorrelation function can be determined by taking the limit in the expression

$$R(\eta) = \frac{1}{M} \cos\left\{\lim_{d\to 0} \frac{1}{d} \cos^{-1}(\cos \eta d \cosh \beta d)\right\}, \qquad \text{(A4–1)}$$

where $\beta = \cosh^{-1} M$ and $d = 1/n$.

Evaluating the limit of this indeterminate form, we obtain

$$R(\eta) = \frac{1}{M} \cos\left\{\lim_{d\to 0} \frac{\eta \sin \eta d - \beta \sinh \beta d}{\sqrt{\sin^2 \eta d - \sinh^2 \beta d}}\right\} = \frac{1}{M} \cos\sqrt{\eta^2 - \beta^2},$$

which corresponds to (27–10).

To determine the power spectra, let us make use of the well-known integral (Bochner [4], p. 72)

$$\frac{2}{\pi}\int_0^\infty \frac{\sin x\sqrt{\eta^2 + z^2}}{\sqrt{\eta^2 + z^2}} \cos \eta\xi \, d\eta = \begin{cases} J_0(z\sqrt{x^2 - \xi^2}) & \text{for} \quad |\xi| \leqslant x, \\ 0 & \text{for} \quad |\xi| > x, \end{cases}$$

where J_0 is the Bessel function of the first kind.

Setting $z = 0$, we have

$$\frac{2}{\pi}\int_0^\infty \frac{\sin x\eta}{\eta} \cos \eta\xi \, d\eta = \begin{cases} 1 & \text{for} \quad |\xi| \leqslant x, \\ 0 & \text{for} \quad |\xi| > x. \end{cases}$$

Combining these equations, we obtain

$$\frac{2}{\pi}\int_0^\infty \left[\frac{\sin x\sqrt{\eta^2+z^2}}{\sqrt{\eta^2+z^2}} - \frac{\sin x\eta}{\eta}\right]\cos\eta\xi\, d\eta$$

$$= \begin{cases} J_0(z\sqrt{x^2-\xi^2}) - 1 & \text{for} \quad |\xi| \leqslant x, \\ 0 & \text{for} \quad |\xi| > x. \end{cases}$$

If $x \neq \xi$, we can differentiate the expression under the integral sign with respect to x. This yields

$$\frac{2}{\pi}\int_0^\infty [\cos x\sqrt{\eta^2+z^2} - \cos x\eta]\cos\eta\xi\, d\eta$$

$$= \begin{cases} \dfrac{xJ_1(z\sqrt{x^2-\xi^2})}{\sqrt{x^2-\xi^2}} & \text{for} \quad |\xi| \leqslant x, \\ 0 & \text{for} \quad |\xi| > x. \end{cases}$$

Setting $x=1$, $z=i\beta$, and comparing the result with the integral (27–14), we arrive (after the corresponding normalization) at (27–15).

Appendix 5

Spherical Functions*

The spheroidal functions were defined in §28 as the eigenfunctions of the integral equation (28–8). Introducing the dimensionless coordinates $\eta = \Omega_m t/2$, $\xi = 2\omega/\Omega_m$ and using the notation accepted for spheroidal functions, this equation has the form

$$\lambda\psi(\eta) = \int_{-c}^{c} \psi(\eta') \frac{\sin(\eta - \eta')}{\pi(\eta - \eta')} d\eta' . \tag{A5–1}$$

In addition, it was shown that the eigenfunction $\psi_i(\eta)$ and its spectrum have the same functional structure, and that the spectrum differs from zero only in the interval $|\xi| \leqslant 1$. This condition can be written more precisely in the following form [39]:**

$$\int_{-\infty}^{\infty} \psi_i(\eta)\, \mathrm{e}^{-j\eta\xi}\, d\eta = \begin{cases} j^{-i}\sqrt{\dfrac{2\pi c}{\lambda_i}}\,\psi_i(c\xi) & \text{for } |\xi| \leqslant 1, \\ 0 & \text{for } |\xi| > 1, \end{cases} \tag{A5–2}$$

where $c = T\Omega_m/4$. The kernel of (A5–1)

$$\varrho(\eta - \eta') = \frac{\sin(\eta - \eta')}{\pi(\eta - \eta')} = \frac{1}{2\pi}\int_{-1}^{1} \mathrm{e}^{j\xi(\eta - \eta')}\, d\xi \tag{A5–3}$$

is symmetric. The eigenvalue λ_i is a part of the energy of the normalized "signal" $\psi_i(\eta)$, which is contained in the "time" interval $(-c, c)$. It follows from (A5–2) that $\psi_i(\eta)$ is an integral, analytical function belonging to the W_1 class; this function does not vanish over any finite interval. Therefore,

* The functions considered here differ from the angular function $S_{0i}(z)$ for the prolate spheroidal coordinates only by the normalization and scale (Morse & Feshbach [50], Part I, p. 642; Part II, pp. 1503, 1576).

** It can be shown that (A5–1) follows from (A5–2): namely, that it is the first iteration of the inverse Fourier transform; therefore (A5–2) is sufficient for determining the function ψ_i.

a finite energy, lower than the total signal energy, is contained in any "time" interval, *i.e.*,*

$$0 < \lambda_i < 1 \quad \text{for} \quad 0 < c < \infty \qquad \lim_{c \to 0} \lambda_i = 0, \qquad \lim_{c \to \infty} \lambda_i = 1. \tag{A5–4}$$

Since the eigenvalues are positive, the spheroidal functions form a complete orthogonal system in the interval $(-c, c)$**; one could use the normalization rule such that

$$\int_{-c}^{c} \psi_i(\eta)\, \psi_j(\eta)\, d\eta = \begin{cases} 0 & \text{for} \quad i \neq j, \\ \lambda_i & \text{for} \quad i = j. \end{cases} \tag{A5–5}$$

It is considered here that the eigenvalues λ_i are not degenerated, *i.e.*, only one eigenfunction ψ_i corresponds to every λ_i ([41], pt. 1).

Any function $f(\eta)$ defined and integrable in the $(-c, c)$ interval, can be represented by the converging Fourier series

$$f(\eta) = \sum_{i=0}^{\infty} a_i \psi_i(\eta), \qquad -c < \eta < c; \tag{A5–6}$$

moreover, from (A5–5),

$$a_i = \frac{1}{\lambda_i} \int_{-c}^{c} f(\eta)\, \psi_i(\eta)\, d\eta. \tag{A5–7}$$

The properties of the above eigenfunctions are not unusual and are characteristic of the majority of symmetric kernels. A peculiar feature of spheroidal functions is that *in addition to being orthogonal in the* $(-c, c)$ *interval, they are also orthogonal in the* $(-\infty, \infty)$ *interval, i.e.,*

$$\int_{-\infty}^{\infty} \psi_i(\eta)\, \psi_j(\eta)\, d\eta = \begin{cases} 0 & \text{for} \quad i \neq j, \\ 1 & \text{for} \quad i = j. \end{cases} \tag{A5–8}$$

In fact, by making use of (A5–3), the following identity can be readily proved:

$$\int_{-\infty}^{\infty} \varrho(\eta - \eta'')\, \varrho(\eta'' - \eta')\, d\eta'' = \varrho(\eta - \eta').$$

* This relationship can be proved directly without resorting to the physical interpretation on the basis of the Bochner-Khinchin theorem [4].

** Here we shall make use of some well-known theorems from the theory of integral equations; a summary of these theorems can be found, for example, in the book of Davenport and Root ([23], p. 371).

Therefore, from (A5–1), we have

$$\int_{\infty}^{\infty} \psi_i(\eta'')\,\psi_j(\eta'')\,d\eta''$$

$$= \frac{1}{\lambda_i \lambda_j} \int_{-\infty}^{\infty} d\eta'' \int_{-c}^{c} \psi_i(\eta')\,\varrho(\eta'' - \eta')\,d\eta' \int_{-c}^{c} \psi_j(\eta)\,\varrho(\eta'' - \eta)\,d\eta$$

$$= \frac{1}{\lambda_i \lambda_j} \int_{-c}^{c} \psi_j(\eta)\,d\eta \int_{-c}^{c} \psi_i(\eta')\,\varrho(\eta - \eta')\,d\eta' = \frac{1}{\lambda_j} \int_{-c}^{c} \psi_i(\eta)\,\psi_j(\eta)\,d\eta\,,$$

and thus the orthogonality condition (A5–5) leads to (A5–8).

For an infinite interval, the system $\{\psi_i\}$ cannot be complete for a class of arbitrary functions, whose square is integrable. Since $\psi_i(\eta)$ belongs to the W_1 class, the expansion in these functions cannot converge (in an infinite interval) to a signal, whose spectrum extends beyond the $|\xi| < 1$ band. However, *a system of spheroidal functions is complete in the W_1 class*. From (A5–2) and (A5–6), the series in the $\psi_1(c\xi)$ functions converges for any spectrum contained in the $-1 < \xi < 1$ band. Therefore, by using the inverse Fourier transform, an analogous series can be obtained for the signal; moreover, this series also converges since the initial series for the spectrum converges. Thus, for any function f belonging to the W_1 class, we have

$$f(\eta) = \sum_{i=0}^{\infty} b_i \psi_i(\eta), \quad -\infty < \eta < \infty; \tag{A5–9}$$

moreover, from (A5–8),

$$b_i = \int_{-\infty}^{\infty} f(\eta)\,\psi_i(\eta)\,d\eta\,. \tag{A5–10}$$

A function belonging to the W_1 class can be represented in two ways: by the series (A5–9) suitable for an infinite interval, or by the series (A5–6) suitable for the $(-c, c)$ interval. Since the expansion in a complete system is unique and both series converge, we also have, considering (A5–7),

$$b_i = a_i = \frac{1}{\lambda_i} \int_{-c}^{c} f(\eta)\,\psi_i(\eta)\,d\eta\,. \tag{A5–10'}$$

This renders a simple method for the extrapolation of functions with a bounded spectrum given in a finite interval ([41], pt. 1). If $f(\eta)$ does not belong to the W_1 class, then the series (A5–9) diverges. However, by making use of the double-orthogonality condition, it is possible to develop an interesting approximation method for the function $f(\eta)$. This method is presented in §29.

The asymptotic behavior of spheroidal functions as $c \to \infty$ will now be considered. Let us introduce new arguments of

$$x = \eta/\sqrt{c}, \quad y = \xi\sqrt{c},$$

and denote

$$\psi_i(\eta) = \psi_i(c, \eta) = \psi_i(c, \sqrt{c}x) = z_i(c, x).$$

Then the expression (A5–2) takes the form

$$\int_{-\infty}^{\infty} z_i(c, x)\, e^{-jxy}\, dx = j^{-i}\sqrt{\frac{2\pi}{\lambda_i}}\, z_i(c, y), \quad |y| < \sqrt{c}.$$

If λ_i does not depend on c, then this equation can only be satisfied in the case where $z_i(c, x)$ has the form

$$z_i(c, x) = A_i(c)\, \varphi_i(x).$$

Therefore, by letting $c \to \infty$ and using (A5–4), we find

$$\int_{-\infty}^{\infty} \varphi_i(x)\, e^{-ixy}\, dx \sim j^{-i}\sqrt{2\pi}\varphi_i(y), \quad -\infty < y < \infty. \tag{A5–11}$$

Thus, as $c \to \infty$, the functions $\varphi_i(x)$ form an orthogonal system in the infinite interval and satisfy (A5–11). As is known, the Hermitian functions satisfy these conditions (Titchmarsh [60], p. 77). Consequently,

$$\varphi_i(x) \sim e^{-\frac{x^2}{2}} H_i(x),$$

where H_i is the Hermitian polynomial.

The factor $A_i(c)$ is defined by the normalization condition (A5–8). Finally,

$$\psi_i(\eta) \sim (\sqrt{\pi c} 2^i i!)^{-\frac{1}{2}}\, e^{-\eta^2/2c} H_i(\eta\sqrt{c}). \tag{A5–12}$$

In particular, for $i=0$, we have

$$\psi_0(\eta) \sim (\pi c)^{-\frac{1}{2}}\, e^{-\eta^2/2c}.$$

An equivalent result was obtained for $\psi_0(\eta)$ by Gurevich ([20], p. 43). Thus, for large values of c, spheroidal functions degenerate into Hermitian functions.

The author does not know of numerical tables for spheroidal functions. Graphs of the first four functions are given in [41], part 1. The eigenvalues $\lambda_0(i)$ to $\lambda_s(c)$ are given also in the same work and in [39] for some values of c. A more detailed table of $\lambda_0(c)$ can be found in [20]. In addition, the expansion of the function $\psi_0(c, \eta)$ in Legendre polynomials suitable for numerical calculations is given in [20].

REFERENCES

1. Bakut, P. A., *et al.*: *Problems of the Statistical Theory in Radar* (in Russian), Vols. I, II, G. P. Tartakovskii, ed., "Sovetskoe Radio", 1963, 1964.
2. Barker, R. H.: Group Synchronizing of Binary Digital Systems, in *Communication Theory*, W. Jackson, ed., London, 1953.
3. Blokhintsev, D. I.: *Quantum Mechanics*, translated from the 3rd and 4th Russian editions by J. B. Sykes and M. J. Kearsley; Dordrecht, D. Reidel, 1964.
4. Bochner, S.: *Lectures on Fourier Integrals*, translated from the German by M. Tenenbaum and H. Pollard; Princeton, Princeton University Press, 1959.
5. Brillouin, L.: *Science and Information Theory*, 2nd ed., New York, Academic Press, 1962.
6. Brown, W. M. and C. J. Palermo: Effects of Phase Errors on the Ambiguity Function, *1963 IEEE International Convention Record*, pt. 4, pp. 118–123.
7. But, E. D.: *Numerical Methods* (in Russian), Fizmatgiz, 1958.
8. Wainstein, L. A. and V. D. Zubakov: *Extraction of Signals from Noise*, translated from the Russian by R. A. Silverman; Inglewood Cliffs, Prentice-Hall, 1962.
9. Vakman, D. E.: *Asymptotic Methods in Linear Radar* (in Russian), "Sovetskoe Radio", 1962.
10. Watson, G. N.: *A Treatise on the Theory of Bessel Functions*, 2nd ed., Cambridge, Cambridge University Press, 1962.
11. Wiener, N.: *The Fourier Integral and Certain of its Applications*, New York, Dover Publications, 1957.
12. Woodward, P. M.: *Probability and Information Theory, with Applications to Radar*, New York, Pergamon Press, 1957.
13. Gonorovskii, I. S.: *Radio Engineering Circuits and Signals* (in Russian), "Sovetskoe Radio", 1964.
14. Goncharov, V. L.: *Theory of Interpolation and Approximation of Functions* (in Russian), GTTI, 1934. 3
15. Gorelik, G. S.: *Oscillations and Waves* (in Russian), Fizmatgiz, 1959.
16. Gorelik, G. S.: Some Microphase Meter Methods in Radio Physics and Optics (in Russian), *Izmeritel'naya Technika*, No. 3, 10 (1955).
17. Gradshtein, I. S. and I. M. Ryzhik: *Tables of Integrals, Series, and Products*, 4th ed., translated from the Russian by Scripta Technica, Inc.; New York, Academic Press, 1965.
18. Green, C. A. and R. T. Moller: The Effect of Normally Distributed Random Phase Errors on Synthetic Gain-Pattern, *IRE Transactions on Military Electronics*, Vol. MIL-6, Apr. 1962, p. 130.
19. Groginsky, H. L. and L. R. Wilson: Effects of Phase Errors on Pulse Compression System, *IRE Transactions on Information Theory*, Vol. IT-8, 1962, p. 321.
20. Gurevich, M. S.: *Radio Signal Spectra* (in Russian), Svyaz'izdat, 1963.
21. Gurevich, M. S.: Signals of Finite Duration, Containing Maximum Share of Energy in a Given Frequency Band (in Russian), *Radiotekhnika i elektronika*, **1**, 3, 313 (1956).

22. Gutkin, L. S.: *Theory of Optimal Radio Reception in the Presence of Fluctuating Noises* (in Russian), Gosenergoizdat, 1961.
23. Davenport, W. B., Jr., and W. L. Root: *An Introduction to the Theory of Random Signals and Noise*, New York, McGraw-Hill Book Co., Inc., 1958.
24. Dirac, P. A. M.: *The Principles of Quantum Mechanics*, 4th ed., Oxford, Clarendon Press, 1958.
25. Dolph, C. L.: A Current Distribution for Broadside Arrays which Optimizes the Relationship between Width and Sidelobe Level, *Proc. IRE*, 1946, Vol. 34, No. 6.
26. Dulevich, V. E.: *Theoretical Fundamentals of Radar* (in Russian), "Sovetskoe Radio", 1964.
27. Zelkin, E. G.: *Design of an Emitting System by the Directional Pattern* (in Russian), Gostekhizdat, 1963.
28. Siebert, W. M.: A Radar Detection Philosophy, *IRE Transactions on Information Theory*, Vol. IT-2, Sept. 1956, pp. 204–221.
29. Ivanova, I. M., Yu. L. Ketkov, and T. S. Yampol'skaya: On the Existence of Barker Codes (in Russian), *Izvestiya vysshei shkoly, radiofizika*, **3**, 5, 911 (1960).
30. Kantorovich, L. V. and V. I. Krylov: *Approximate Methods of Higher Analysis*, translated from the Russian by C. D. Benster; New York, Interscience, 1958.
31. Kartianu, G.: *Frequency Modulation*, Publishing House of the Academy of Sciences of the Rumanian People's Republic, 1961.
32. Key, E. L., E. N. Fowle, and R. D. Haggarty: A Method of Designing Signals of Large Time-Bandwidth Product, *1961 IRE International Convention Record*, pt. 4, pp. 146–154.
33. Klauder, J. R.: The Design of Radar Signals Having Both High Range Resolution and High Velocity Resolution, *The Bell System Technical Journal*, **39**, 4, 809–820 (1960).
34. Klauder, J. R., A. C. Price, S. Darlington, and W. J. Albersheim: The Theory and Design of Chirp Radars, *The Bell System Technical Journal*, **39**, 4, 745–808 (1960).
35. Cramér, H.: *Mathematical Methods of Statistics*, Princeton, Princeton University Press, 1946.
36. Courant, P. and D. Hilbert: *Methods of Mathematical Physics*, Vol. I, New York, Interscience, 1953.
37. Cook, C. E.: Transmitter Phase Modulation and Pulse Compression Waveform Distortion, *The Microwave Journal*, **6**, 5, 63–69 (1963).
38. Cook, C. E.: Effects of Phase-Modulation Errors on Pulse Compression Signals, *1962 IRE International Convention Record*, pt. 4, pp. 174–184.
39. Landgrebe, D. A. and G. R. Cooper: Two-Dimensional Signal Representation Using Prolate Spheroidal Functions, *IEEE Transactions on Communication and Electronics*, Vol. 82, 1963, pp. 30–40.
40. Lacy, P.: Automated Measurements of Phase and Transmission Characteristics of Microwave Amplifiers, *1963 IEEE International Convention Record*, pt. 3, pp. 119–121.

41. Landau, H. J., H. O. Pollak, and D. Slepian: Prolate Spheroidal Wave Functions, pts I, II, III, *The Bell System Technical Journal*, **40**, 1, 43–64 (1961); **40**, 1, 65–84 (1961); **41**, 4, 1295–1336 (1962).
42. Levin, B. P.: *Theory of Random Processes and its Application in Radio Engineering* (in Russian), "Sovetskoe Radio", 1960.
43. Lezin, Yu. S.: *Optimal Filters and Integrators for Pulse Signals* (in Russian), "Sovetskoe Radio", 1963.
44. Lerner, R. M.: Signals with Uniform Ambiguity Functions, *1958 IRE National Convention Record*, pt. 4, p. 27.
45. Leipnik, R.: Entropy and Uncertainty Principle, pts. I, II, *Information and Control*, **2**, 64–79 (1959); **3**, 18–25 (1960).
46. Van der Maas, G. I.: A Simplified Calculation for Dolph-Tchebyscheff Arrays, *Journ. of Appl. Phys.*, **25**, 1, 121 (1954).
47. Mayo, B. R., P. W. Howells, and W. B. Adams: Generalized Linear Radar Analysis, *The Microwave Journal*, **4**, 8, 79–84 (1961).
48. Middleton, D.: *Introduction to Statistical Communication Theory*, New York, McGraw-Hill Book Co., Inc., 1960.
49. Minkovich, B. M.: Concerning the Quasi-Optimal Linear Center-Fed Antennas with Continuous Current Distribution (in Russian), *Radiotekhnika i elektronika*, **4**, 6, 1671 (1959).
50. Morse, F. M. and H. Feshbach: *Methods of Theoretical Physics*, pts. I, II, New York, McGraw-Hill Book Co., Inc., 1953.
51. Ramp, H. O. and E. R. Wingrove: Principle of Pulse Compression, *IRE Transactions on Military Electronics*, Vol. MIL-5, Apr. 1961, p. 109.
52. Root, W. L.: Radar Resolution of Closely Spaced Targets, *IRE Transactions on Military Electronics*, Vol. MIL-6, April 1962, No. 2, p. 197.
53. Sokolov, I. F. and D. E. Vakman: Optimal Linear Broadside Arrays with Continuous Distribution of Current (in Russian), *Radiotekhnika i elektronika*, **3**, 1, 46 (1958).
54. Sproull, R. L.: *Modern Physics; a Text for Engineers*, New York, John Wiley & Sons, 1956.
55. Storer, J. E. and R. Turin: Optimum Finite Code Groups, *Proc. IRE*, Vol. 46, No. 9, 1958, p. 1644.
56. Sussman, S. M.: Least-Square Synthesis of Radar Ambiguity Functions, *IRE Transactions on Information Theory*, Vol. IT-8, No. 3, 1962.
57. Tartakovskii, L. B.: On the Theory of a Double-Curvature Mirror (in Russian), *Radiotekhnika i elektronika*, **4**, 11, 1821 (1959).
58. Tartakovskii, L. B. and V. K. Tikhonova: Synthesis of Linear Emitter with a Given Amplitude Distribution (in Russian), *Radiotekhnika i elektronika*, **4**, 12, 2016 (1959).
59. Temes, C. L.: Sidelobe Suppression in a Range-Channel Pulse-Compression Radar, *IRE Transactions on Military Electronics*, Vol. MIL-6, April 1962, No. 2, p. 162.
60. Titchmarsh, E. C.: *The Theory of Fourier Integrals*, London, Oxford University Press, 1948.

61. Turin, R. and J. Storer: On Binary Sequences, *Proc. Am. Math. Soc.*, **12**, 3, 394 (1961).
62. Urkowitz, H., C. A. Hauer, and J. F. Koval: Generalized Resolution in Radar Systems, *Proc. IRE*, Vol. 50, Oct. 1962, pp. 2093–2105.
63. Fal'kovich, S. E.: *Reception of Radar Signals on the Background of Fluctuating Interferences* (in Russian), "Sovetskoe Radio", 1961.
64. Fel'd, Ya. N. and L. D. Bakhrakh: Present State of Antenna Synthesis Theory, *Radiotekhnika i elektronika*, **8**, 2, 187–205 (1963), English translation in *Radio Engineering and Electronics Physics*, **8**, 2, 163–179 (1963).
65. Kámpe de Fériet, J., *et al.*: *Functions of Mathematical Physics.*
66. Fradin, A. Z.: *Microwave Antennas*, translated from the Russian by M. Nadler; R. C. Glass, ed.; New York, Pergamon Press, 1961.
67. Kharkevich, A. A.: *Outlines of the General Communication Theory* (in Russian), Gostekhizdat, 1955.
68. Kharkevich, A. A.: *Spectra and Analysis* (in Russian), Gostekhizdat, 1953.
69. Hoffman, D. A.: Synthesis of Linear Coding Circuits, in *Lectures on Communication System Theory*, E. J. Baghdady, ed.; New York, McGraw-Hill Book Co., Inc., 1961.
70. Helstrom, C. W.: *Statistical Theory of Signal Detection*, New York, Pergamon Press, 1960.
71. Khurgin, Ya. I. and V. P. Yakovlev: *Methods of the Theory of Integral Functions in Radar Physics, Communication Theory, and Optics* (in Russian), Fizmatgiz, 1962.
72. Chandrasekhar, S.: *Stochastic Problems in Physics and Astronomy.* English language edition, *Selected Papers on Noise and Stochastic Processes*, Nelson Wax, ed.; New York, Dover Publications, 1954.
73. Shirman, Ya. D.: A Method of Improving the Resolution of Radar Stations and Arrangement for its Realization (in Russian); Author's Certificate No. 146803 registered under No. 46/974/40 on 25 July 1956, *Byulleten' Izobretenii*, No. 9, 1962.
74. Shirman, Ya. D.: Stochastic Analysis of Optimal Resolution (in Russian), *Radiotekhnika i elektronika*, **6**, No. 8, 1961.
75. Shirman, Ya. D. and V. N. Golikov: *Fundamentals of the Theory of Detection of Radar Signals and the Measurement of Their Parameters* (in Russian), "Sovetskoe Radio", 1963.
76. Shifrin, Ya. S.: The Statistics of the Field of a Linear Antenna (in Russian), *Radiotekhnika i elektronika*, **8**, 3, 400–407 (1963); English translation in *Radio Engineering and Electronics Physics*, **8**, 3, 351–358 (1963).
77. Zakai, M.: A Class of Definition of "Duration" (or "Uncertainty") and the Associated Uncertainty Relations, *Information and Control*, **3**, 2, 101–115 (1960).

Author Index

Subject Index

MIX
Papier aus verantwortungsvollen Quellen
Paper from responsible sources
FSC® C105338

If you have any concerns about our products,
you can contact us on
ProductSafety@springernature.com

In case Publisher is established outside the EU,
the EU authorized representative is:
Springer Nature Customer Service Center GmbH
Europaplatz 3, 69115 Heidelberg, Germany

Printed by Libri Plureos GmbH
in Hamburg, Germany